Black Hole
Survival Guide

Black Hole
Survival Guide

Janna Levin

Alfred A. Knopf · New York · 2020

THIS IS A BORZOI BOOK
PUBLISHED BY ALFRED A. KNOPF

Copyright © 2020 by Janna Levin

Artwork copyright © 2020 by Lia Halloran

All rights reserved. Published in the United States by
Alfred A. Knopf, a division of Penguin Random House
LLC, New York, and distributed in Canada by Penguin
Random House Canada Limited, Toronto.

www.aaknopf.com

Knopf, Borzoi Books, and the colophon are registered
trademarks of Penguin Random House LLC.

Library of Congress Cataloging-in-Publication Data
Names: Levin, Janna, author.
Title: Black hole survival guide / Janna Levin.
Description: New York : Alfred A. Knopf, [2020]
Identifiers: LCCN 2020015159 (print) |
LCCN 2020015160 (ebook) | ISBN 9780525658221
(hardcover) | ISBN 9780525658238 (ebook)
Subjects: LCSH: Black holes (Astronomy)—Popular
works.
Classification: LCC QB843.B55 L48 2 (print) |
LCC QB843.B55 (ebook) | DDC 523.8/875—dc23
LC record available at https://lccn.loc.gov/2020015159
LC ebook record available at https://lccn.loc.gov/
2020015160

Jacket art by Lia Halloran
Jacket design by John Gall

Artwork by Lia Halloran

Manufactured in Canada
First Edition

Contents

Black Hole
Survival Guide

CHAPTER 1

Entrance

Black holes are nothing.

Black holes are special because there's nothing there. There is no thing there.

I probably accepted black holes whole, as complete conceptual entities, before I was able to doubt, before I had intuition to combat. They were fodder for fantasy. I received the fact of their existence without resistance. Gullible and without prejudice, I could see the plausibility, appreciate the curiosity, their peculiarity, accept the universe as it was presented. Maybe it was the same for you. It's very unlikely that this is your first encounter with the astrophysical oddity that is a black hole, the warp in spacetime so strong that not even light can escape.

I don't know what it was like where you were, but I had a pretty unspectacular view, astronomically speaking, out of my childhood bedroom. I would edge to the foot of the bed and strain to look at the sky through the window frame, a portal to a green yard below, the seams between neighboring plots demarcated by shrubs or trees, patchwork real estate scooped together under a rounded cap of atmosphere. The ground would darken first, then the trees, but the dome would hang on to the vaguest splattering of light for the longest. Ours was not an urban burnished sky, but still light pollution diluted the view that I stared into night after night. I never expected the vista to improve or expected anything new, only the usual indistinct few bright speckles, like rain stains on a windshield.

I don't remember when the feeling started, the first time that longing tugged at me, but before I even recognized the desire, I pined to know what else was out there, like a restless dog pacing at the door. I wanted to be free of ordinary confines. To fly there and explore. Frustrated by the fact of the heaviness of my feet on the Earth, striding at the base of the sky, restless to be let in. I wonder how many of us inherited this longing, millennium after millennium, generation after generation, child after child bound to the crust, rapt at the illusion of a ceiling, compelled to crack through and defy our puniness and our limitations.

As a kid, I never imagined I would become a scientist. If you had told me I would become a physicist, I'm sure I would have been offended. Scientists build bombs and memorize equations. I'm not sure if I acquired that cliché or if the stereotype was one of my own invention—that scientists could not be creative, or just weren't out of stubbornness. I didn't understand the profound freedom of constrained creativity, visionary originality that explodes in the face of fundamental limits.

Limits have incited revolutions. The limit of the speed of light hinted at relativity. Einstein fantasized about riding a light beam and imagined that time stood still. He gave up the absolutism of an inflexible space and time in favor of the absolutism of the speed of light. That concession forced us to reconceive of our universe as an impermanent, living process with an origin in a big bang, still expanding with the energy of creation, and home to extravagances like black holes.

The quantum revolution was incited in parallel by the limit imposed by Heisenberg's uncertainty principle, which asserts that particles as we thought we knew them do not exist. We are pressured to reimagine the fundamental nature of reality as an uncertain fog of possibilities, of particles that are there and not there. The pressure to yield to this seemingly terrible constraint guided us toward revelations that would otherwise have been unattainable. We rewrote reality in an astonishing new language. We have discovered quarks, photons, neutrinos, condensed matter, dead neutron stars, the Higgs boson, superconductors, and quantum computers in the most precisely tested paradigm in all of physics.

The computer revolution too was incited by a limit, an insurmountable limit to mathematical knowledge. The incompleteness theorems, which proved that there were mathematical facts that could never be proven, led Alan Turing to dream of artificial intelligence and biological machines. Turing proved that most facts about numbers are unknowable: there is an infinite list of irrational numbers with an infinite list of unpredictable digits. His musings led him to imagine a machine that one day will think and to the epiphany that, indeed, *we* are machines who think.

The severity of the constraints imposed by physical law and mathematical precision does not squash creativity. The limits are the scaffolding enabling creativity. Limits can be worthy adversaries that galvanize our best, most inventive, most agile natures. Before I succumbed to the seduction of the elegance and transcendence of limits, I did not understand the thrill of imagination crashing into truth.

As a student, not yet on a trajectory to become a scientist, I was nostalgic for that longing to leap into the atmospheric pool of blue midnight high out my window. I realized I missed that sensation, the urgent desire to leave the planet. Wherever I was by then, I was still under the same sky. From a slightly different perspective I trod at the foot of that blue dome, that overly lit blanched blue sky, and wanted in. Deterred by reality (we've never gone farther than the Moon), I redirected my focus down to the page, following mathematics to places our bodies couldn't go. Mathematics alone cannot tell us what specifically is out there in our universe. The mathematics can speculate only about what is possible. And sometimes mathematics allows us to explore pure potential before any physical manifestations of that potential are discovered.

Black holes were like that, a purely mathematical construct on the page, benign in virtual form, in typescript on paper, unverified for decades, unaccepted for decades, absurd, maligned and denied by some great geniuses of the twentieth century, until physical evidence of real black holes in the galaxy was discovered. Find one just a few thousand light-years away—a light-year being the distance light can travel in a year, nearly 10 trillion kilometers, the distance you would travel driving at the average highway speed limit for 10 million years. Take a left at that yellow star and veer toward that star cluster. Wandering at the base of the sky, we are under them. We are above them. Black holes in their abiding darkness are scattered plentifully among the stars, which themselves are scattered plentifully, like somber glitter infiltrating the void. We are in orbit around one in the center of our Milky Way galaxy. We are pulled toward another in the Andromeda galaxy.

I want to influence your perception of black holes, to shuck away the husk a bit, get closer to their darkest selves, to marvel at their peculiarity and their prodigious character. We can take a road less traveled, follow a series of simple observations that culminate in an intuitive impression of the objects of our attention, which are not objects at all, not things in the conventional sense.

A friend takes me out in New York to discuss the essentials to include in a black hole survival guide. An accomplished science writer, he asks me to clarify: "Don't I already know everything about black holes?"

"Do you know they are nothing?"

He looks at me unblinking for a long time, considering. He pauses to toss some salted peanuts into his mouth and while chewing he says, "I guess I don't know anything about black holes." Then slightly bemused at this realization, we eat the rest of the bar snacks and drink our wine while we talk about things more familiar.

CHAPTER 2

Space

Weightlessness and Free Fall

Black holes are much maligned, depicted unfairly as behemoths when they are often benign and actually by nature quite small. Still, before you travel, you should do your research and consider the hazards. The perils, in fairness, are exceptional and unmatched if you are not careful. As with nature untamed here on Earth, black holes demand respect for safe navigation. After all, you are a trespasser on their territory.

Black holes were the unwanted product of the plasticity of space and time, grotesque and extreme deformations, grim instabilities. Honestly, they're not such an anathema to scientists now. Black holes are a gift, both physically and theoretically. They are detectable on the farthest reaches of the observable universe. They anchor galaxies, providing a center for our own galactic pinwheel and possibly every other island of stars. And theoretically, they provide a laboratory for the exploration of the farthest reaches of the mind. Black holes are the ideal fantasy scape on which to play out thought experiments that target the core truths about the cosmos.

When in pursuit of a black hole, you are not looking for a material object. A black hole can masquerade as an object, but it is really a place, a place in space and time. Better: a black hole *is* a spacetime.

Imagine an empty universe. You have never seen or experienced such a pristine place, a vast nothingness that is the same everywhere—vast and bare. And flat—still three-dimensional, but everywhere flat.

Here is the sense in which an empty universe is a flat space: if you were in flat space you would float on a straight line. Despite the fact that you are not falling in the colloquial sense, free motion is called free fall. You are in free fall as long as you do not fire any rockets, get pulled or pushed—essentially as long as there is only gravity. Just surrender to space. If free motion can be traced by straight lines, and lines that begin parallel never cross, then the geometry of space is flat.

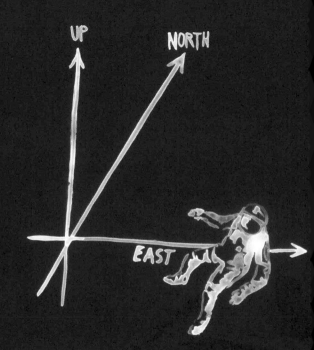

Chances are surpassingly bad that you are in free fall right now. Chances are also surpassingly bad that you are in a flat space, since there is no such place anywhere in our galaxy. Sit on a chair and you are not in free fall. The chair pushes on you to stop your fall. Stand on a floor and you are not in free fall. The floor pushes against your feet to prevent your plummet to street level. Lying in bed we feel heavy. We say gravity pulls us down. But we have it all wrong. Totally inverted. What you feel is not gravity but rather the atoms in the mattress pushing against your atoms. If only the bed would get out of your way, and the floor, and all the lower floors, you would fall, and falling is the purest uninterrupted experience of gravity. Only in the fight against gravity do you feel its pull, an inertia, a resistance, a heaviness. Give in to gravity, and the feeling of a force disappears.

The classic setting for the idea of free fall is an elevator. You are high up in an apartment building in an elevator. You feel a force against your feet. That force is between your feet and the elevator floor and keeps you in the cab. It's a force between matter. Now, if you are interested in pure gravity, without the interference of interactions between matter, you must get rid of the elevator cab somehow. So you enlist some-one to cut the cable. The elevator falls and you with it. During the descent, since you and the floor fall at the same rate, you would float in the elevator. You don't fall to the floor, because the floor is falling too. You can push off the walls and tumble in the air. You seem weightless, as though you were an astronaut in the space sta-tion. You can pour water out of your water bottle and drink the droplets from the air, like astronauts do. You can release a pen, a phone, a rock in front of you, and these float too. Einstein called this profoundly simple observation—that we experience weightlessness when we fall—the happiest thought of his life.

The spoiler is that your atoms interact with the atoms in the Earth's surface, and that would ensure an unhappy end to your free fall when you hit the ground. But that's not gravity's fault. The shattering of your bones would be due to other forces, like forces between atoms. (If you were made of dark matter, you could fall right through the Earth's crust and sail on down.)

We can't run the elevator experiment for long before the Earth gets in the way. So instead imag-ine you float far from the Earth, far from the Milky Way. Imagine a fictitious empty space, except for you and your space suit. If you sent tracers, threw a projectile in each of the three spatial directions, those projectiles would free-fall. Now imagine each projectile leaves a helpful trail illuminating its path; soon, a grid of straight lines would be visible. You could see plainly that space was flat—free-falling objects follow straight lines—and that space was empty, except for you and your space suit and your tracers and the luminous trails charting the grid. Gravity is so weak that none of these little pieces have a noticeable impact on the flat emptiness of it all.

The universe is not empty. We are very aware that we are bound to the Earth. The Earth is bound to the Sun and the Sun to the Milky Way galaxy. The Milky Way is bound to the neighboring galaxy Andromeda, both residing in the Virgo supercluster of galaxies. And the Virgo supercluster senses all the other galaxies and all the accumulated energy in our observable universe. So we don't live in a flat, empty spacetime.

Astronauts also don't float in empty space. They can see the Earth spin and the Sun roll along. They are falling and weightless, but on a path we've been accustomed to calling an orbit, an orbit around the Earth in orbit around the Sun in a glacially long orbit around the galaxy. Their

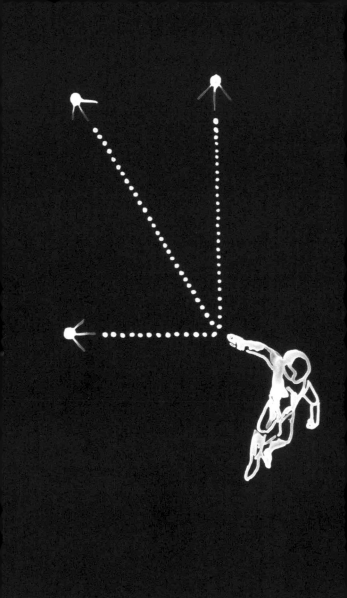

paths aren't straight. Their paths are curved into a circle around the Earth sewn into the circle around the Sun sewn into the path around the galaxy, because free-fall paths are curved when the sky isn't empty. Because space is curved by the presence of matter and energy.

Curved Space

You can prove that you live in a curved space and not a flat space from the comfort of your couch by throwing things. Throw something and watch the arc it traces. The projectile will not travel in a straight line. The path paints a curve in space, an arc. All the objects we throw follow curved paths toward the Earth. We could travel around the globe throwing projectiles and all objects we throw from international couches will bow to the ground. We could document the results and draw a three-dimensional grid of the curved paths and thereby construct a map of the shape of the space around the Earth. The lesson: the Earth deforms the shape of space. And you can map that shape by drawing free-fall paths.

Free fall depends on the speed at which you toss something around the planet. Drop a wrench to the Earth and it follows a line straight down.

Throw the wrench across the room, and the descent is along an arc, the same arc a car would descend along if thrown at the same speed and in the same direction as the wrench. Throw the wrench faster and the arc gets longer. Throw the wrench fast enough and it will clear the curve of the Earth and launch into orbit. Throw the wrench faster still and it will float away from the Earth forever until caught on the curve of another celestial object, like Jupiter or the Sun, and tumble on a different path.

Planets fall around the Sun. No engine pushes them. They trace ovals, always clearing the atmosphere. The Earth falls freely around the Sun, and the Moon falls freely around the Earth.

We put lots of human-made objects into orbits, which are just free-fall paths. Once the launched spacecraft get where they need to be, the engines are turned off and they can fall forever in orbit around the Sun or, more commonly, the Earth. Some mission specialists fight against the inclusion of thrusters in the spacecraft designs, in case a decline in funding encourages the space agency to maneuver the satellite out of orbit and incinerate the bounty in the atmosphere. Defunct satellites haunt space, ghostly litter in orbit for the lifetime of the solar system.

The International Space Station falls freely around the Earth. The astronauts in the space station float because they are falling like the ill-fated elevator occupants, not because they don't feel the gravitational effect of the Earth. They do. The station is only a few hundred kilometers high and very much under the Earth's influence. The gravitational effect is traced by curves in the shape of space, and the ISS follows one such natural curve, an ever-falling circular orbit. The astronauts and the space station travel nearly 28,000 kilometers per hour to complete a full circle every 90 minutes, out of the sunlight and into the Earth's shadow and out again into the Sun's radiance. They move fast enough that they always clear the Earth's atmosphere while falling and thereby never crash into the surface.

From Einstein's happiest thought (we fall in space weightlessly), we deduce with the most ordinary observations (tossed projectiles from the vantage of our Earth-covering couches) that free-fall paths are curves in space. Gravitation *is* curved spacetime. And that insight was Einstein's greatest.

Einstein's unabashed devotion to simplicity shows in the childlike wondrousness of his unembellished, spare thought experiments. He knew he did not understand standard-issue grade-school gravity—how can the Earth pull on the Moon, when they are not touching? Einstein did not understand gravity and neither did anyone else, but other scientists of his time either didn't appreciate the severity of the failing or they didn't pause to consider the implications. But because he did not understand gravity and he admitted as much, Einstein challenged the most accepted and elementary aspects of reality.

Before Einstein, it was customary to consider gravity to be a force of one body acting on another, but a force that mysteriously did not require actual contact. After Einstein, the language changed. Gravity was cast as a curved spacetime. How can the Earth pull on the Moon without touching it? It doesn't. It doesn't pull on the Moon at all. It exerts no force. Instead, the Earth bends space. And the Moon tumbles freely.

Black Holes Are a Space

Far from a black hole, the curves in space are the same species as those around the Sun or the Moon or the Earth. If the Sun were replaced by a black hole tomorrow, our orbit would be unchanged. The curve we would fall along around a black hole sun is nearly identical to the curve we fall along around the actual Sun. Of course, the perpetual dusk would be apocalyptically cold and dark. But our orbit would be just fine.

The Earth is on average about 150 million kilometers from the Sun, which is about 1.4 million kilometers across. In comparison, a black hole the mass of the Sun would be 6 kilometers across. You can approach much closer to a black hole and remain unharmed than you can to the Sun, despite a black hole's reputation for voraciously consuming all and sundry. You only really notice a radical difference between the space around a black hole and the space around the Sun when you get within a few hundred kilometers of the center of each. And you can't get that close to the Sun without incinerating.

Bore inside the solar plasma and the gravitational pull of the Sun eases off. As you approach the center, you leave behind some of the Sun's mass. The curving of space inside the Sun's atmosphere becomes more gradual as the mass beneath you diminishes.

By contrast, no matter how close you approach to a black hole, the source never diminishes. The curving only gets sharper. Black holes are special because you leave none of the black hole's mass behind you; it is as though all of the mass is concentrated ahead of you. Always. You can approach infinitesimally close to the center of a black hole and still feel all that mass in front of you.

Keep a safe distance from an unobtrusive black hole and you will neither be torn apart nor sucked up. Black holes just are not the catastrophic engines of destruction they're portrayed to be, at least not until you veer recklessly close, not until you cross the point of no return, and then admittedly circumstances can get harrowing. Even if you approach boldly close, within several widths of the black hole, you could set up your space station, shut off the engines, fall freely in a stable orbit that takes mere hours to complete, and enjoy the scenery for as long as supplies last.

CHAPTER 3

Horizon

Light Falls Too

You should appreciate the hazards of encountering a black hole unawares. A black hole is invisible in the absence of any tracers, just darkness against darkness. You may well not realize the threat before your fate is secured. You must carry a powerful light source to reveal in backlight the clandestine black hole, an unilluminated disk, an absence in a bright world.

Light lives in space too and has to follow some path. If you shine a flashlight from your couch, you don't notice the light beam falling toward the Earth. The lines appear straight. But they're not perfectly straight. The comparative inflexibility of light's route through space is attributable to its intrinsic speed. Light only travels at one speed, the speed of light. The free-fall curves for light—launched as light must be, at the cosmic speed limit—are straighter than the curves of slower objects. So the bend that the Earth's gravity imparts to the path of light is more subtle, straighter, and more difficult to detect.

The bending of light offered the first test of general relativity. On the twenty-ninth of May, 1919, the Moon eclipsed the Sun to allow a thin ray of light from the Hyades star cluster to fall into Arthur Eddington's telescopes. With the blinding solar rays occulted, the faint image of Hyades could be collected. But at the time of the eclipse, Hyades was positioned directly behind the Sun from the Earth's perspective. If light traveled along straight lines, none of the luminosity from Hyades should have made it to the Earth. The cluster would spray its rays in all directions, and those headed toward the Earth would fall into the Sun. If light traveled along the curved paths predicted by relativity, then the Sun would act like a lens and bend an image of Hyades our way.

The Sun was eclipsed in totality only from the point of view of a swatch of the Earth that cut through Brazil at dawn, across the Atlantic Ocean as the Earth turned, and fell off Africa by dusk. From any given vantage point between sunrise and sunset, totality would last less than seven minutes. Barely six months after the end of the First World War, on the small island of Principe off the coast of Africa, Eddington and his team waited out heavy rains and cloud cover until the Sun broke through, the eclipse already under way. Although behind the Sun, the star cluster was visible to their telescopes (glossing over some ambiguities in the data), proving that light did not travel on straight lines around our star, but rather curved ones that deflected the light toward Africa.

Eddington presented his analysis of the resultant photographic plates from the eclipse expedition in Principe, in conjunction with those from his team near the equator in Brazil, and his announcement made headlines, catapulting Einstein's fame in the English-speaking world. Eddington became the conduit that transmitted these discoveries, discoveries that transcended political animosity and nationalism stirred up by the war. Eddington, unlike some of his compatriots, had no urge to denigrate Einstein's accomplishments, though he was a citizen of a country so recently a bitter enemy. Einstein was born in Germany, Eddington in England. Out of the shadow of the war into the shadow of the

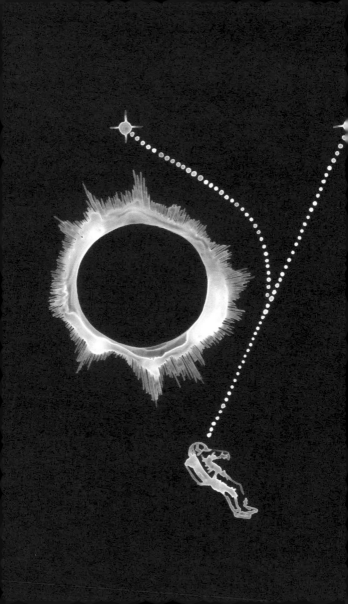

Moon, they were citizens of the same Earth, and the theory of relativity was heralded as one of humanity's greatest achievements.

Eddington measured the slight deflection of light's path around the Sun. Outside every black hole, there is a curve in space so sharp that light can fall around the hole in a circular orbit. You could jet-pack to the location of light's circular orbit and hover there. You will need to fire engines to resist your fall into the hole. Once there, you could shine a light on your face. Your face will reflect some of the light (if we didn't reflect light, we'd be invisible) and send the image of your face in a reverse orbit toward the back of your head. You could wait a few tenths of a millisecond for that light to reflect off the back of your head then circle back about and land in your eyes. You could watch your own back.

When you get near enough to a black hole, the curves are sharper and the free fall faster. It becomes harder to stop your fall. You will need a considerable payload of fuel to accelerate you to sufficient speeds. To escape from the Earth, a rocket launched from Cape Canaveral must reach more than 11 kilometers per second. The escape velocity from the Moon is about 2.5 kilo-

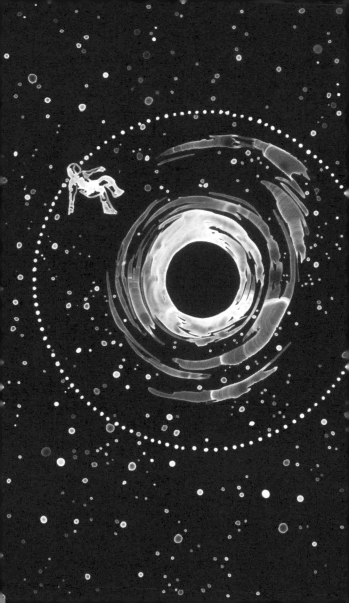

meters per second, the Moon being less substantial than the Earth. The escape velocity from the Sun is more than 600 kilometers per second, the speed of the plumes of plasma able to blow off the hot stellar atmosphere into the solar system in the form of ionized winds.

The nearer you approach a black hole, the faster you need to drive your spacecraft to resist falling. In the emptiness that surrounds a black hole, you can get closer and closer to the center until you hit a proximity outside the hole at which no amount of jet thrusters is sufficient to reverse your decline. You would need to reach a velocity greater than the speed of light to escape, a speed of almost 300,000 kilometers per second. And since nothing can travel faster than the speed of light, you would fail to escape. Extrication impossible, you plunge irrevocably.

That special location where the escape velocity climbs to the speed of light defines the infamous event horizon: "The region beyond which not even light can escape." I'm using unattributed quotes because everyone seems to use that exact sentence. We all say it at some point in our lives: "The region beyond which not even light can escape." I said almost these exact words at the opening of this guide.

Light shining from the precise location of the event horizon seems to hover there, traveling at its fixed speed and still unable to escape, a fish swimming against an insurmountable waterfall of space. Knock the light slightly and it loses the battle, falling down the hole.

The event horizon is the extent of the black hole's shadow. Anything that crosses the event horizon is forever lost to the outside. From the exterior, the black hole is dark. It is black. It's a black hole. If you were to fall through that shadow, you'd find no physical, material surface. You'd just plummet in the dark emptiness, the moment of transition unspectacular, as unspectacular as stepping into the shadow of a tree.

The essence of a black hole is the event horizon, the severe demarcation between events and their causal relationships. Events interior to the horizon can have no effect on the outside, can tolerate no transmission of any kind to the exterior of the horizon, although the reverse is not the case. Happenings outside the horizon can be transmitted to the interior. The world beyond the black hole can have consequences for the world within. The black hole is thoroughly opaque from the outside, utterly transparent from the inside.

Should the choice arise, you are advised to fall into as big a black hole as possible. If you are very small compared to the black hole, you will hardly notice that the horizon is curved, like you hardly notice that the Earth's surface is curved. If you stood on a basketball, the curvature under your feet is more noticeable than the curvature of the entire Earth. The bigger the black hole, the less you'll notice. Your feet and your head could more or less fall in unison. You'd find yourself compromised if the hole was so small that your feet fall across the horizon along their own path that diverges from that of your head. Your connective tissue would have to resist the fall of your feet away from your crown and eventually the ligaments would fail and rip and the consequences would be pretty gruesome. But traverse the event horizon of a big black hole, no problem.

So you could survive the fall across the shadow, although you'd never find your way out again. And once inside the black hole you will eventually get pulverized, a caution we'll detail later.

If you fall close to the center of the Sun, you could in principle escape the gravitational pull, but you'd be incinerated in the nuclear furnace of the core, a scenario that concludes with equally emphatic finality. At least you can get a million kilometers closer to the center of a black hole and live, in principle.

Your ruination inside a black hole the mass of the Sun typically takes less than microseconds. You could prolong your life expectancy to as much as a year in a black hole trillions of times more massive. You should aim for a bigger black hole if you intend to survive long enough to reflect on your experience. You may not want to prolong your demise if you cannot tolerate the existential dread, the gnawing suspicions, the cogent predictions about your unstoppable departure from this adventure.

If you find yourself approaching an utterly dark shadow, visible only in contrast to a bright background of light, beware. Avoid at all costs. Maintain a safe distance. If you get too close you will need all the fuel in the universe to escape, and that will not be enough. It will be hard to assess if the shadow is small and nearby or big and far away. If you do cross the shadow, you will be the first human to be dismembered by spacetime and obliterated near the infamous singularity, prophesized to be an actual hole in the center of the black hole, a tear in the fabric

of spacetime that leads nowhere. Take solace in your observations, but they will be yours alone. None of your transmissions will make it out of the hole to home. All documentation of your demise will follow you into oblivion.

CHAPTER 4

Nothing

The black hole is a not a thing. It is nothing.

A bare black hole is pure empty spacetime—no atoms, light, strings, or particles of any kind, dark or bright. It's empty space—or, in physics slang, the vacuum.

You've probably already heard a description of the formation of black holes that goes something like this: crush matter down to tremendous densities and you'll make a black hole. This is true. Compressed matter is one avenue to the creation of black holes. Heavy stars collapse under their own weight at the end of their life cycle and the material of the star crushes to catastrophic densities until a black hole forms. Still, stellar collapse is not the only imaginable mechanism for their formation. The dense quash of matter is often mistakenly taken as synonymous with the black hole. But that's not the essence of the black hole. Black holes are not stuff.

Here's how this works. Imagine a star collapses. Spacetime is curved due to the hot sphere of matter. The curvature gets stronger just outside the star as the star compresses. The denser the matter becomes, the stronger the curves impressed just outside the shrinking outer surface. A splash of plasma must launch faster and faster to escape the star until the star becomes so dense that the speed to escape reaches the speed of light. An event horizon forms and nothing from the star, not even the light it shines, can ever escape again.

Now the star continues to collapse, marking the event horizon indelibly on the shape of spacetime. The star continues to fall, the implosion unstoppable. The material that formed the event horizon is gone, leaving empty space—and an event horizon—in its wake.

Effectively, the event horizon *is* the black hole. The event horizon occludes the interior. There is no fact about the interior of the black hole that can possibly be transmitted to the outside universe because the event horizon forbids the transmission of any information on any facts about the interior. The actual matter of the star becomes irrelevant, except for the impression left behind in the form of the event horizon.

Shed the impression of the black hole as a dense crush of matter. Accept the black hole as a bare event horizon, a curved empty spacetime, a sparse vacuity that hypnotized me into this peculiar occupation.

I've tried to peel away the veneer to expose black holes at their barest, their most fundamental. A glorious void, an empty venue, an extreme, spare stage, markedly austere but, yes, able to support big drama when the stage is occupied. Black holes are a place in space and they barricade their secrets. They are a place that can also behave like an object. They are empty but have mass.

When I talk about the mass of a black hole and then say there is no stuff there, that seems like a sleight of hand. If there is no stuff, it would be fair to ask, how can it have any mass? The original star might be gone, but the energy equivalent of the mass of the star was imparted to the black hole. In the case of stellar collapse, physical things made of real material and having conventional mass impress their heft on space, leaving behind a gravitational attraction of equivalent value.

If we discover a means by which the imploded object could form a remnant, an unrecognizable vestige of crushed matter, deep inside the event horizon, then we might rely on convention and connect the mass of the black hole to actual extant material. But this interpretation would be misleading, would allow the challenge to go unnoticed. The black hole is not the remnant, even if it hides one. Since the event horizon obscures from the outside the fate of the infalling matter, the black hole is indistinguishable whether the matter is destroyed, survives as a remnant, or finds life in another universe.

We can therefore say black holes have mass, even if all the stuff that went into their formation is gone. And black holes act quite a bit like objects with mass. Black holes themselves can fall like massive objects along the curved paths of other heavy objects. Black holes can orbit stars and galaxies and other black holes. They can be pulled and pushed, and it will be harder to do so the more inertia they have—that is, the more mass they have. So we describe a black hole in terms of its mass, even if the original heavy object that may have made the black hole was siphoned out of existence in the interior.

The acclaimed American relativist John Archibald Wheeler said in a lecture in 1967, "[The star] like the Cheshire cat fades from view. One leaves behind only its grin, the other, only its gravitational attraction. . . . Light and particles . . . go down the black hole." And with that quote, Wheeler embedded the term "black hole" in the physicists' lexicon, some fifty years after the original discovery of the mathematical description.

So this is what you must remember: The black hole event horizon is empty. The black hole is no thing. The black hole is nothing.

Imagine you are an astronaut alone in space. No planets in view, no spacecraft. No distant stars. No source of light. Imagine the latent terror, the quiet of space, the strange sensation of floating, the unspeakable dark between the wealth of stars.

If you do transit the event horizon of a big black hole, the trip will be undramatic. You should feel no pain. You won't crash into any surface. Aside from the menacing darkness, the transit across the event horizon should be perfectly comfortable. You would cross the heavy shadow and the nothingness you see inside would look indistinguishable from the nothingness you saw outside. You would be oblivious to

a demarcation of inside or outside. Without light to map the terrain, the disorientation would be thorough and impenetrable. You could fall inside the hole and survive the transition, very briefly unaware of the grim prospects for the future. Nothing is the worst thing you could encounter. Beware the event horizon. The void is inescapable once crossed.

CHAPTER 5

Time

I've casually referred to curved space as a proxy for curved spacetime. While black holes deform space around them—their radically warped interiors cloaked behind an event horizon—black holes also deform time.

Consider this experiment. You are in a dark, sealed elevator cab. The cable was just cut. You don't know where you are or how you got there. You decide to test the world around you. Your experience of the world will be ordinary. You float as though in empty space. Any clocks you carry run as anticipated. You might suspect for a moment that you are falling freely in a flat space. But if you could open a window and look out to see other observers at some distance in other falling elevators, you might discover that you and the others are converging

toward a center, and thereby realize that actually space is curved, perhaps around a black hole. You might also notice that your clocks are not in sync with some of the others' clocks.

On comparison, your measures of space and time disagree with those made by some of the other observers traveling along different curves. Why? Because measures of space and measures of time are relative.

Principle of Relativity

A very old principle of relativity has served as a guide to expose physical laws. This principle suggests that unless nature has a compelling reason to the contrary, the laws of nature should look the same to observers in free fall. It's a simple principle that has proven tried and true. If the principle ever falters as a sensible signpost to the truth, the consequences will be remarkable. But so far, the principle hasn't failed architects of physics.

Here is the simplest example of a principle of relativity. Left is relative. If we face each other, my left is your right. So which way is left? The laws of nature cannot possibly depend on which direction you call left. It would be ridiculous if laws were different on the left than on the right, if atoms weighed more on the left than on the right. Who's left? Who's the preferred person who chose left so auspiciously? If there is no fundamental principle for picking out one person's left as preferred, then nature should not prefer one person's left over any other person's designation of left.

Galileo thought about relativity, but we have the benefit of space travel to make our thought experiments more transparent. Galileo contemplated relativity on Earth, which is so much more confusing. Einstein contemplated relativity in moving trains. Again, so much more confusing. In space, we can take away more landmarks. We can take away *all* landmarks, and that is to our advantage.

Imagine you are alone in an empty universe with no stars, no planets, no spacecraft. Doesn't matter how you got there. It's just you and only you. Given your sense of orientation, you can ask, Which way is up? Which way is right? Since there is no physical reason to single out a given direction, you should not hang too much on your declarations. They are yours to declare but also arbitrary. Nature cannot possibly care which way is up.

Now try to determine if you are moving. The

same answer is now slightly more unnerving, almost existential. How can you know? Are you drifting slowly? Are you coursing near the speed of light? There is no way to know. There is no air to breeze against your skin. You bring your atmosphere with you in your helmet. There is actually no physical experiment you can perform that will reveal to you that you are in motion. It's not that you can't uncover the answer. It's more that there isn't an answer to this question. Or, put more strongly, the question is meaningless. Absolute motion is as arbitrary as left.

Imagine up another astronaut, Alice. Let Alice drift upside down relative to you, our previously solitary astronaut, who watches her pass right on by. Suppose her orientation is backward relative to you. You are upside down relative to her. You are backward relative to her. Neither of you should find a particular challenge in the idea that you have assigned opposite directions to up and opposite directions to left. Neither of you should find any value in one choice over any other. But then you have to accommodate a fairly bigger concept. Which one of you is in motion? You could say she is moving. She could say you are moving. And there is no physical law that can select either astronaut as preferred. No

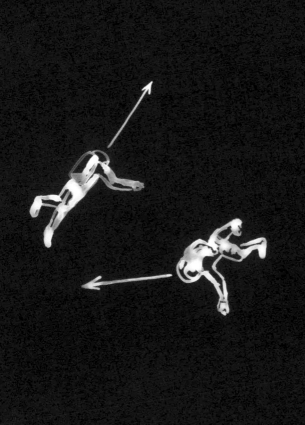

one is preferred. No one is moving. No one is pointing up. No one is facing left.

Galileo recognized that motion is relative. More precisely, as long as motion is completely uniform and follows free-fall paths in our empty imaginary universe, there is no way to determine absolute motion because there is no absolute motion. If instead of traveling uniformly, you are lurching about, say by firing jet packs, then both astronauts can agree that you are the one lurching and you can both agree that the lurching was caused by the accelerations imparted by thrusts from the jet pack. Uniform motion, however—the free fall along straight paths in empty space—is as relative as the designation east. You and Alice agree you are in relative motion. Still, there are no absolute terms that can assign the motion unambiguously to one of you over the other.

The Speed of Light

You and Alice do everything you can to determine your absolute motion and you fail. You fail because nature, by all evidence, respects this beautiful principle of relativity. If there is no compelling criterion to prefer you over Alice, then nature does not. No one is preferred.

In adherence to this principle, the laws of physics, in the most abstract sense of the fundamental codes of nature and in the more concrete sense of the mathematizations we uncover, have the same form for you and for Alice. You say projectiles travel on straight lines. She says projectiles travel on straight lines. You observe that for every action there is an equal and opposite reaction. She quotes the same rule. You measure the strength of chemical bonds and diffusion rates of solutions in a beaker and her experiments toss up the same numbers. You accept the relativity of left and right, the relativity of your motion, in exchange for agreement on a short list of facts that constitute the laws of physics.

This was all good in the minds of physicists for many centuries. Then one fact precipitated directly from the laws of physics to demand that we accept a much more profound relativity. The fact that threw down the challenge is the rule of the constancy of the speed of light.

Light is a form of oscillating electrical and magnetic energy, otherwise known as electromagnetic radiation. The speed at which the radiation propels forward is a fundamental number. fixed in the very laws of electricity and of magnetism. The speed of light is a fact of nature. A constant, universal fact. That speed, designated by the letter c, equals nearly 300,000 kilometers per second.

Since motion is relative, it is fair to ask: Light's speed is c relative to what or to whom? The answer: Light's speed is always c relative to everything and everybody.

The relative speed of a basketball is not an immutable fact. A player can hold the ball stationary or propel the ball toward the basket. An astronaut's relative speed is not a fact. You might see Alice drift past slowly. Or you might watch her cruise by quickly. Or even not at all. You could be lost together, floating side by side forever. But light cannot move slower or quicker; it will always whiz by at the same universal speed, regardless of who is observing.

Now you and Alice have a truly beautiful concept to grapple over. You measure the speed of light. She measures the speed of light. You get the same number, c. You watch each other's experiments and disagree with every other determination of every other velocity of every other thing. You release a basketball just as she passes. She says it's moving away from her quickly, as quickly as you are. You disagree. You see the basketball as motionless, just hovering in front of you.

She throws atoms together in an accelerator and says they move fast. You say some of her smashed atoms move faster, some slower, depending on whether they move toward you or away from you. She holds a lightbulb and says it's at rest; you say no, it's moving, along with her. You are both perfectly content that these pedestrian observations differ precisely as anticipated based on your different points of view due to your relative motion.

But she says, as you course past, "Look at the light shining from my lightbulb, it travels at speed c." You say, "Yes, quite right."

You should be amazed.

Nature has no criteria to prefer you over Alice. So nature chooses no one. For whom is light's speed c? Einstein answered: everybody. Everyone has to measure the speed of light to be c.

Time Is Relative Too

To orient yourself, you might declare up to be the direction of your head, west the direction of your left arm, north the direction you face. And you could well conclude with perfect consistency that you are not moving in space. You haven't budged up or down, east or west, north or south. You are at rest in space as far as you are concerned. But you also know, no matter what, that you are not at rest in time. Time flows past you—or, with better imagery for our picture, you flow through time.

Now a distinct speck resolves into a companion, Alice, who moves relative to you, possibly even nearly as fast as light. And you perceive that she moves through space. She moved up to down. She of course says, no, she is at rest in space in her assignment of up and down. Fine. You've gotten over this. But here's the intriguing thing, the hint at something very profound to come. She says she is at rest in her map of space but that she is moving in time. She doesn't deny she is older than she was when you first came into focus on her skyline, that she can count her breaths and they accumulate and that there are more to come. According to her, she moves only in time, in her time. If you were to draw her path through your spacetime, you would say, "You have moved through my space and my time." And she would say the same of you. It's as though you are rotated relative to each other not just in space but in spacetime.

YOUR TIME

ALICE'S TIME

ALICE'S SPACE

YOUR SPACE

This picture could be a wrong picture. Who says we can draw a map with time on it? But if it's a correct picture to draw, if there can be a map not just of space but of spacetime, then the analogy is compelling. The map suggests that her time is not elapsing at the same pace as yours. It looks from the drawing as though part of what she calls time you call space. It looks like you will say she hasn't experienced the same time as you have because the duration she traveled in her time direction is not the same as the duration she traveled in yours. And, of course, she could draw her map and accuse you of the same. Even time is relative.

Einstein realized that both you and Alice will measure the same value for the speed of light if your times are rotated relative to each other *and* your distances are rotated relative to each other. The greater your relative motion, the greater the rotation between you.

The spacetime diagram is both right and wrong. It rightly gives the same speed of light for everyone. You see light travel a distance through space equal to 300,000 kilometers according to your rulers in a duration of time of one second according to your clock. Despite her entirely different labeling of spacetime, Alice also sees the light travel 300,000 kilometers according to her rulers in a duration of time of one second on her clock. You both measure the same law of nature, but at the expense of a shared experience of time and space.

The flaw in the spacetime diagram is the geometry imposed. The diagram is drawn on a flat sheet of paper in space on your guidebook, and that limitation creates a misrepresentation. A familiar example of a map that requires new rules for reading off distances and shapes is an ordinary geographic map of the Earth. We know when we look at the spherical surface of the Earth projected onto a flat sheet of paper that distances and sizes are distorted by the projection. We have taken a curved surface and falsely flattened it. We can use the flat map perfectly well as long as we read carefully and apply the proper rules for adding distances and interpreting angles and that sort of thing.

The empty spacetime in which you and Alice float can also be drawn on a flat sheet of paper. But the drawing introduces a distortion analogous to the distortion introduced when the spherical Earth is projected onto a flat sheet of paper. Just as the Earth should be more accurately drawn on a sphere, spacetime should be drawn on a different kind of surface, one for which the rules for measuring distances are not the familiar rules we apply to paths drawn on a sheet of paper. The spacetime surface can be mathematized formally (it's called Minkowski spacetime) but can't be so easily drawn. For now, we project onto our sheet of paper and learn the rules of the projection (which I won't provide here, since that would require an elaborate digression). We draw these spacetime diagrams often. I personally love them. We just don't read them in the conventional way.

Black Holes Dilate Time

Unlike flat spacetime, which has no landmarks to measure against, a black hole creates a definite landmark to leverage against. Now you and Alice can say unambiguously whether or not either of you moves relative to the event horizon.

Suppose you and Alice are in a space station far from a black hole. You see the shadow through your telescope. You decide to investigate alone and leave the space station and Alice. You sever your tether and drift toward the hole. You and Alice agree to send each other messages at regular intervals. You begin to fall freely along a path directed toward the event horizon. At regular intervals you fire your jet pack so that you can pause motionless relative to the hole and relative to Alice, who maintains the space station at a safe distance. You realize that the closer you get to the shadow, the more you have to fire your jet packs in order to stay motionless relative to that shadow.

The black hole is dark. You can make out the shadow because the 300 billion stars bulking up the galaxy light the world around the black hole. The black hole warps space into a lens, re-fracting the image of the galaxy. All light that veers too close is lost in the event horizon, which paints a solid dark circle on your sky.

As you advance progressively closer to the hole, the time displayed on your clocks seems normal to you, the readings corresponding to your intuitive experience of the passage of time. But Alice complains that your clocks appear to run slowly, that the time on your clocks is lag-ging behind the time on her clocks. You cannot disagree. Her clocks are indeed further ahead of yours. And Alice is aging more rapidly than you. Her music in the background plays at an absurdly quick tempo, and movies on the screens behind her seem to fast-forward at an uninter-pretable pace.

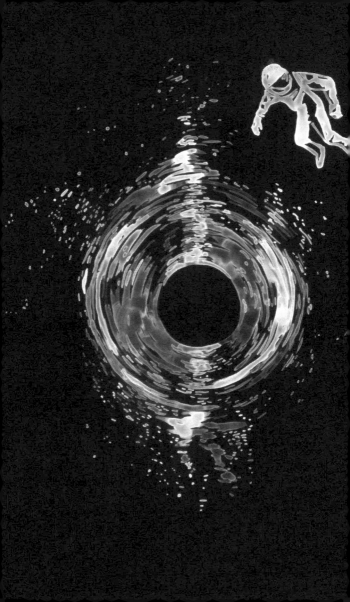

During the pauses as you fire your rockets to halt your fall, you are not in relative motion, not with each other and not with the hole. Still, your measures of space and time are rotated relative to Alice's by a bit more with each advance you make toward the event horizon, because black holes warp time as well as space. Your time will run slower than hers until you reach the event horizon, when your clocks appear to Alice to stop, your passage of time inert.

Since the escape velocity at the event horizon is the speed of light, your signal to Alice, encoded on light as all interspace signals tend to be, will try to race outward but will be unable to escape. As you cross the event horizon, you will pass by the light you put there, the light that sits still even though it is moving at the speed of light. It is as though space is gushing like a waterfall into the hole and the light is struggling at its prescribed speed limit, still c, but making no progress against the waterfall of space. Light gets stuck at the horizon, while you are swept down into the hole with the gushing space. From far away, it will appear as though time on the horizon stands still, as though all clocks are frozen, including biological clocks, as though your crossing of the event horizon takes an infinite time.

But as far as you are concerned, your time is perfectly normal, nothing about the clock you built has changed, the mechanism is perfectly sound, and your transition across the event horizon is uneventful. All is well, though you only have handfuls of instants to live before you face an unknown fate in the interior.

Before you die, you will see that Alice's clocks are ticking quickly, running much faster than yours. The video messages she sends you are a blur as she races through her life. She dies of old age before you die in the center of the black hole. In the fraction of a second before you confront your own extinction, you watch your space station deteriorate from exposure to interstellar elements, winds off stars, lethal bursts from supernovae, and colliding dead stars.

Time is dilated, or in more words: time as measured by clocks that fall into the black hole runs slowly relative to time as measured by clocks far away. And this dilation extends until time seems to stop altogether at the event horizon, relative to clocks far away.

If time and space are measured according to the clocks infinitely far out, the sentinel of those clocks will be forced to conclude that inside the event horizon what they once called time is now space and what they once called space is now time. Think in analogy of rotating left into right. How can time get any slower than not progressing at all? It is as though time rotated entirely into space and space into time.

Inside the event horizon, to move outward you would have to do the impossible and travel faster than the speed of light. Since nothing can travel faster than light, all viable paths point inward. The future points toward the center. You are doomed. You can only move forward toward the singularity, which exists in your future. The event horizon exists only in your past. You fall into that future as inexorably as the companion you left behind moved forward in time toward her future: death by old age, nations conquered and fallen, entire civilizations come and gone.

The black hole singularity that we naïvely think of as at the center of a sphere is really at a future point in time and not a point in space at all. You cannot see the center of the black hole. Light can no more travel toward you from the singularity than light can travel into the past. There is no turning back time. Death in the singularity is in your future.

CHAPTER 6

TARDIS

Your jet packs are useless at the event horizon. You turn them off and succumb to the fall. The black hole was a solid disk of darkness from outside. But once you cross to the interior of the event horizon, you discover that you can still see out. You are not plunged in darkness. The event horizon poses no obstacle to the galactic luminosity raining down. Light drops right through the horizon, harbinger of distorted images. The black hole is dark on the outside, but it can be bright on the inside.

Through the one-way window of the horizon, you view the universe beyond. While there's nothing you could do to halt your slide, you could for those brief moments look through the pane and watch the universe evolve. Light from the galaxy that floods through the event

horizon portrays a radically sped-up version of thousands or millions or billions of Earth years. The light that spills into your eyes plays out the downfall of civilizations or paparazzi flashes of eons of exploding stars. As you fall, the hole's throat narrows and seizes all of the transmitted luminescence into a focused bright whiteness. Like in a near-death experience, you see the light at the end of the tunnel. Only it's a total death experience.

If we follow the mathematics to the brutal end, the general theory of relativity predicts that the black hole's interior cinches, the spacetime curving calamitously and unmitigated, to form a singularity where all paths terminate. A singularity might as well be a cut in spacetime. The material of the original star flies into that rip and is blotted out of existence. Not only is the actual imploded substance behind the event horizon irrelevant to the structure of the black hole, it's gone.

Once across the horizon, I abandon my defense of black holes. You plummet inevitably into that singularity. It will be a rocky ride as your matter perturbs the environment and the spacetime undulates. As you fall toward the singularity, you are badly broken. The part of

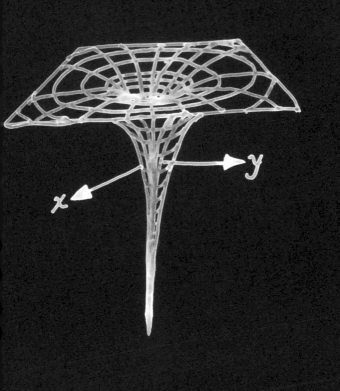

your body closest to the singularity is accelerated drastically faster than the part of your body farthest from the singularity, stretching you miserably. Simultaneously, your overall anatomy is forced to converge toward that point, crushing you. In a microsecond, less time than it would take to blink your eye, you are simultaneously flayed, shredded, and pulverized to death. Your organic matter is then pummeled, tattered, and inevitably shattered into elementary constituents. Ultimately, your fundamental bits spray toward the cut in spacetime and cease to be.

The rip leads to nowhere. The singularity is an end to space and time, an end to existence. There's no future ahead once a thing gets crushed and pushed through the singularity. Death by singularity is the paramount existential death — the death of your fundamental particles, the removal from reality of you and your constituent stuff. Actual nonexistence.

Yet we don't have to resign ourselves to this travesty or to the inevitability of singularities. Singularities, as they involve malefic infinities, deserve to be treated with great suspicion. They are such an anathema to the entire paradigm of the scientific pursuit of reality that essentially all physicists suspect general relativity ceases to be the complete physical description of gravity at such dramatic scales, the singular core a false prophecy. Rephrased: the mathematics is telling us that the physical description offered by relativity is broken there. General relativity cannot be the whole story precisely because it predicts the singularity. The alternative to abandoning faith in relativity is much worse: the existence of singularities would mean the physical universe is deeply and pathologically misbehaved.

Maybe in the black hole abyss, instead of a singularity there is some kind of leftover from all that infalling material, a quantum remnant at the catastrophically high energies and curvatures of the very center of the hole. It's conceivable that all of the matter that created the hole and subsequently fell into the hole is trapped in an as yet unknown state of quantum matter, that all of the subatomic particles—previously healthy constituents of a star, matter we think we understand pretty well—instead compress into less space than a trillion trillion times smaller than one hydrogen nucleus.

It's also conceivable that it's nonsense. There aren't all that many supporters for the remnant hypothesis. As long as we're succumbing to speculation, here's a favorite: once inside of the black hole, maybe everything ruptures into a white hole, something like a new big bang into another part of the universe, because black holes can be bigger on the inside than they are on the outside. Like Doctor Who's TARDIS. There could be a whole other universe inside that hole.

A black hole is a place, a location in spacetime, eerily dark and bare and empty. And yet, scientists have not been able to answer the seemingly simple question, Where do we go if we fall in? The mystery of the black hole interior imposed by the event horizon gives black holes a special cultural aura not permitted most astrophysical phenomena.

Regardless, you get crushed to death long before a remnant forms or a big bang occurs. Revising the singularity into something more sensible won't save you. You will be mangled and granulated, but your scraps could be part of a greater ecosystem. If your splinters are not blotted out of existence through a singularity, if your scraps could wait in the form of a jittering quantum remnant in the core of the black hole, incidental rubble will spill down with you, more errant space junk, mixing with any other debris accrued. The remnant persists indefinitely, a slim hope of a possible future. Or your elements will be shared with a new universe, blown out into a big bang, reordered into generations of stars, some of them recycled finally into a microbial life-form on a new piece of dirt destined to fall into another black hole.

CHAPTER 7

Perfection

Black holes are astounding. Not only are black holes dead stars, not only do they cannibalize neighboring stars, not only will they eventually absorb the entire galaxy and one another (after a terribly long time), not only do they ignite the most powerful engines in the known universe, not only are black holes vortices of slop and mayhem—they're perfect.

Black Holes Have No Hair

Black holes are perfect. By that I really mean that black holes are featureless. Try to put a blemish on a black hole. It won't stick. The hole will shake off any imperfections and settle down to its perfect, featureless self. Regardless of how unique the original star (let it be composed entirely of donkeys) and regardless of how collapse is initiated (I can barely think of something crazy enough; maybe the donkeys huddle together until there are enough of them that they implode under their own weight), the end result will be the same, a featureless black hole, exactly like every other perfect black hole originating from a crushed stellar atmosphere or pulverized diamonds or squashed antimatter sludge, nuclear waste, photons, or fridge magnets.

Try your worst. Deform a black hole as much as conceivably possible. Collide two black holes. In the end you will make one black hole and that black hole will be every bit as perfect as every other black hole. As the two event horizons merge, the spacetime near the two holes heaves and surges in stormy waves. Just as a struck bell will send out sound waves until the vibrations cease, the spacetime sends out waves in the curves in spacetime—gravitational waves—and rings down until quiet. That ring down marks the final throes of the colliding pair. The gravitational waves carry away all the marks and blemishes, and a quiet, perfect, spinning black hole remains.

Throw in a star, a mountain, a goat. The black hole event horizon will deform slightly as the spacetime adjusts, the additional mass building up the black hole, but that deformation will ring off quickly, which is to say that gravitational waves transport away the deformation, leaving the outside horizon as smooth and flawless as ever.

The only distinguishable features of a black hole to an outside observer are its mass, its electromagnetic charge, and its spin. All black holes of a given mass, charge, and spin are identical to all other black holes of that mass, charge, and spin. Those three identifying characteristics completely determine the geometry of the black hole spacetime—that is, the size and shape of the event horizon and the surrounding spacetime.

"Black holes have no hair," as John Wheeler quipped. If you could deduce any other features of the interior, any features other than mass, charge, and spin, it would be as though lines of information were emanating from the black hole, as though the black hole had hair. But the event horizon forbids the flow of information outward and therefore forbids the black hole from acquiring hair. "Black holes have no hair," at least not for long. Any hair you try to give them will either fall in or be radiated away, restoring the hole to a pristine form. And so the black hole will remain featureless and without defect.

Black Holes Are Macroscopic Fundamental Particles

After a while, we on the outside have no way of knowing what went into the black hole, since no information gets to those of us on this side of the event horizon, precisely because black holes are featureless. Consequently—and this is beautiful—from out here, all black holes of a certain mass (and spin and charge) are identical to all other black holes of that same exact mass (and spin and charge). They must be, because how could we possibly tell them apart? The event horizon equalizes all black holes so that from outside, you cannot tell the difference between a black hole formed from pure light and those formed from bars of gold, or feathers, or radioactive uranium, or copies of *The Idiot*. From the outside at least, all black holes of a given size (and spin and charge) are identical and no experiment can distinguish between them.

Black holes are perfect in the same spirit that a subatomic particle like an electron is perfect. Find an electron and it is identical to every other electron in the universe. Electrons are entirely interchangeable, unlike things with electrons in them, things like people or field guides.

Black holes are in some sense fundamental

particles of gravity. And there are no others like them in the universe that we know of, except other black holes. All other ideal particles are subatomic. So maybe there are subatomic black holes too. Unstable, possibly unthinkably heavy as subatomic particles go, but tiny objects we might make at a particle accelerator.

Mini black holes could be created if we slam together particles so that their energy is concentrated in a small enough volume to reach black hole proportions. The smallest fundamental black hole particle would weigh about 22 micrograms theoretically, about a few thousandths the weight of a sesame seed, but would be about 100 million trillion trillion times smaller. Twenty-two micrograms might not seem very heavy, but the microscopic black hole is about 10 million trillion times heavier than a proton and hundreds of millions of trillions of times smaller. In your kitchen, you could easily compile collections of particles that are in the 22 microgram range, like a tiny pile of flour. But the pile of flour is made up of countless smaller particles and is very diffuse, not at all dense. It's easy to make big heavy things, like buildings and spacecraft. It's hard to make tiny heavy things. The smaller, the heavier, the harder to produce. Mini black holes are the heaviest, smallest fundamental particles conjectured.

The Large Hadron Collider in Switzerland smashes beams of particles in a narrow 27 kilometer circular accelerator highway and watches the debris spray toward a series of detectors located along the ring. It is currently the largest, most powerful particle accelerator ever built and may hold this title for eternity. The LHC reaches peak energies 10 million billion times too low to make even a quantum black hole. Still, there was some legitimate research that hypothesized the existence of lighter quantum black holes, ones the LHC might produce, if our universe hides extra spatial dimensions beyond the three we observe daily. But that's a digression.

Physicists publicly exchange these ideas, casually referring to the LHC as a potential black hole factory. Some people overhear the conversations and panic ensues. Lawsuits are filed in an attempt to impose an injunction against a flick of the LHC switch. Independent reviews from scientists deem the machine safe for use, the switch is flicked, and the LHC finds the Higgs, better known as the Goddamn particle, appropriated in the popular lexicon as the God particle. Interestingly, the argument was not that the LHC was guaranteed black hole free. Rather, the statement was that any mini black holes wouldn't destroy the world. First smash particles to create a deformed black hole, which then radiates all the hair, all of the imperfections, to settle down to a perfect fundamental black hole. That black hole then swiftly decays through a quantum process, as will be explained later in this guide, into a blast of Hawking radiation.

If you do aspire to create your own microscopic black holes, your prospects for survival would benefit if you made the smallest black holes, as they are the most unstable to emission of Hawking radiation and will vaporize without too much carnage. Tiny black holes would be harmless to the fate of the world, too short-lived to assimilate the instruments, the tunnel, the scientists, the entire facility, Switzerland, the Earth. If you do make a small black hole, consider throwing in some charged matter. Then you can capture the charged hole with magnetic tweezers and confine the precious merchandise to a hot box, where a delicate and unsustainable equilibrium balances the tendency for the black hole to grow against its tendency to decay. Leaving you to it. Will be a tense lab.

Not everyone appreciates the confidence of

theoretical physicists: Let's make a black hole factory; what could go wrong? Here's a more prosaic but important argument. Cosmic rays strike the Earth's atmosphere with energies beyond the range the collider reaches, and neither our planet nor any others have disappeared down a mini black hole. If microscopic black holes are made in the Earth's atmosphere, and we suspect they are not, they have been too unstable to induce the Earth's destruction.

Sadly, no black holes were detected at the LHC. The only mini black hole factory that seems viable is the big bang. Spacetime expands out of the big bang in a hot, high-energy event, 10 million billion times more energetic than the LHC. Microscopic black holes could have formed within the earliest moments in the lifetime of the universe. Primordial black holes also would disintegrate into other debris through Hawking radiation and would thereby have disappeared long ago, contributing some of the raw material for the formation of the first generation of stars to die and implode into large, long-lived astrophysical black holes.

There's no evidence for these primordial mini black holes. But there is compelling evidence for black holes as heavy as stars, formed in due course, when the universe was able to produce them after a few billion years. So that's where we're heading. We have to leave this austere terrain of the purely theoretical, the pure nothingness of spacetime. Bare isolated black holes still hide from us. But occasionally they are not alone, and they can tear off hunks of nearby material and throw the matter around to reveal themselves, let us know they're there, like an invisible man playing in the snow.

CHAPTER 8

Astrophysics

I need to admit now that the black hole, which is abstractly a place, begins to act very much like a thing in the actual universe. Telescopes operated by satellite or balloon or even laid on the ground can collect light and conclude that black holes are real. So we have to move black holes out of the airy terrain of abstract theory into reality and also move black holes from utterly dark to blindingly bright. You will have to confront jets of matter and antimatter beamed across light-years, even millions of light-years, demolished stars, clumps of material splattering around the hole at nefarious speeds. The darkest astrophysical object in the universe, a veritable hole in space that emits no light, a black hole is transformed ironically into the engine for a light source that outshines any other in the universe.

We have seen quasars, the entire core of an ancient galaxy shining energetically enough for us to see some billions of light-years distant. Supermassive black holes millions or billions of times the mass of the Sun drag galactic driftwood—entire stars, gas and debris, denizens of the astronomical galactic nucleus, the ephemera of the conglomeration's formation—into a hot mess tumbling down into oblivion. Material is caught in an electromagnetic squall driven by the black hole. The matter renders the invisible visible, tracing the storm like dirt in a tornado. The black hole can spin up the miasma into a luminous jet, which is propelled millions of light-years, a staggering beacon we can see far into the observable universe.

When they were first spotted by terrestrial technology, they were called quasi-stellar radio objects and later quasars when their extragalactic origin became apparent. They looked bright and small, like stars, but were scattered outside of the plane of the galaxy, which was a hint that quasars didn't actually live here in the Milky Way. They are billions of light-years away, which means they are old, the light traveling all that way for all that time to get here, and rare, which means the universe doesn't make them as often anymore.

Quasars, occupying the cores of their galaxies, are a kind of active galactic nucleus powered by supermassive black holes. With the mass of millions or billions of Suns concentrated within a region smaller than our solar system, an active galactic nucleus is a heavy anchor, accumulating a dense and populous center. There could be tens of thousands of smaller black holes and other dead stars and some live stars orbiting the nucleus. The supermassive black hole may have its own origins in the seeds of dead stars, in stellar-mass black holes that collided and merged and grew to be the elephantine core of the galaxy, although nobody yet really knows how they formed or got so heavy. Maybe they were not formed from dead stars but maybe instead they directly collapsed out of primordial material in a younger universe. However they arose, there are as many supermassive black holes as there are galaxies, hundreds of billions in the observable universe.

There may have been hundreds of collisions and mergers between entire galaxies. Galaxies will cut right through each other, the stars too sparse to actually physically strike other stars. Mostly interstellar gas collides as the galaxies swing through each other, gravitationally disrupting entire solar systems to send them whirling. The galaxies fly apart and then slump back together for another pass and on and on until the two blend together inseparably. These collisions, which themselves took billions of years to resolve, fan dust and stars into the dense centers. The spray of galactic material swirls into a disk, which washes down the black hole sink. The galactic nucleus ignites. Matter continues to splatter onto the spinning disk of accumulated debris, the accretion disk, hot and radiant. So much energy is released by the accreting matter that the region surrounding the black hole becomes thousands of times brighter than the brightness of all the stars in the galaxy.

Magnetic fields twisted by the central engine— the supermassive black hole—lend wires for charged particles to slide along, tossing matter back out and away at nearly the speed of light, and the energy radiated by those relativistic particles escapes along tight beams, thin powerful jets that pierce intergalactic space across a range of millions of light-years.

An active galactic nucleus will shine brilliantly, luminous enough for us to see from billions of light-years away, a billion years in the future, until the disk of accreting material is spent and the black hole goes dormant and dark.

If somehow this little guide, this relic of some thoughts, endures a myriad of existential threats, the advice to relay to future space travelers is that you ought to stay clear of black hole jets. Consider these jets to be an astronomically augmented, black hole–powered ray gun. The jets will accelerate particles to the highest energies conceivable, producing lethal X-rays and gamma rays. A direct blast of the jet could burn off the protective atmosphere of entire planets, boiling their cores, ensuring the extinction of all indigenous life-forms. The most powerful jets from supermassive black holes blow craters in neighboring galaxies, exterminating any species that may be evolving on billions of planets there. Take precautions. Stay clear of the line of fire. If even lightly irradiated by the fire hose of high-energy particles from a small black hole, you would suffer DNA damage and the anticipated degradation of healthy cells. Intense exposure to radiation will damage your central nervous system, altering motor and cognitive functions, denying you the mental capacity to take evasive action. Radiation sickness would descend as electrons are stripped from atoms, breaking chemical bonds and damaging your body's tissues. That's your best-case scenario: being woefully scorched despite protective gear and precautions and miscalculated guidelines for safe distancing. With a direct hit, anticipate vaporization.

Avoid the jets until the violent era ends. Long ago, our own galaxy may once have had all of the properties of a quasar. Eventually the supermassive black hole at the center runs low on consumables and that active phase of the galaxy's evolution quiets down. The jets eventually turn off, the galactic center remains bright but not blindingly so, and the denizens of the galactic core orbit safely and only very rarely fall into the hole. The darkness permitted by the core's quiescence allows for a recently emerged species to see beyond our own galactic neighborhood across the 13.8 billion years since the big bang.

The supermassive black hole in the heart of our own Milky Way is still there. Sagittarius A* (pronounced "Sagittarius A-star") is so named since from our perspective the galactic center lies beyond the constellation Sagittarius. Astronomers monitor Sag A*, as the name is often shortened, from extreme peaks on the Earth's surface. The Andes slice off a narrow strip of

Chilean desert, the driest brittle Earth topped with the most arid atmosphere. Turbulent winds dissipate against the peaks so that the Atacama Desert to the west roasts in mostly idle air. The quiet atmospheric conditions are deeply dark at night, largely protected from light pollution, so the Atacama appeals to astronomers stranded on Earth. On top of crumbly secondary mountain ranges, a whole series of international observatories perch above a stony terrain laced with the occasional frothy salt lake.

Things must fall into Sag A* from time to time, littler black holes and entire stars, but the era of active accretion is long over. We know the supermassive black hole is still there because some remarkably patient observers spent two decades cautiously tracking the orbits of stars and from the simplest rules of gravitational attraction deduced the mass and size of the otherwise invisible object at the focus of the orbits and concluded: big and heavy—it's a black hole. Just as skidding rocks trace the shape of a hill, the motions of the stars trace the curves in space-time. One star in particular executes a full orbit in sixteen Earth years, traveling around a dark focal point faster than several thousand kilometers per second at its closest approach, which is a few times greater than Neptune's closest approach to the Sun. From these motions, the presence of a supermassive black hole is deduced and estimated to possess a mass 4 million times the mass of the Sun.

In a few billion years, when we collide with our neighbor galaxy Andromeda, the tidal pull of the crash will kick up galactic dust, which will then spill into the galactic center and probably reignite our own supermassive black hole. But in the meantime, the Milky Way's center is congested but fairly quiet. No quasar, no jets.

With patience, astronomers might eventually see bright matter slosh onto a black hole, but that's not quite the same as seeing the black hole directly. We have indirectly inferred the presence of black holes when they've collapsed from the remains of dying stars, cannibalized companion stars, powered quasars and jets, and captured stars in their orbit. We have even heard black holes collide and merge, shaking spacetime like mallets on a drum.

An image of the event horizon's shadow would be closest to an actual picture of the black hole. A black hole against the dark backdrop of empty space is truly invisible. Still, even quiescent black holes often maintain debris in disks hot enough to helpfully cast the shadow of their event horizons. Even light directly behind a black hole gets redirected our way, so the disk should appear to surround the black hole, allowing for a luminous contrast against which its shadow is visible.

Snapping an image of a black hole is impeded by their diminutive size. Black holes are tiny, despite their dramatic reputation as weapons of mayhem and destruction. A black hole the mass of the Sun would have an event horizon a mere 6 kilometers across. Compare that to the 1.4 million kilometer breadth of the Sun itself. Sag A*, almost 26,000 light-years away, is more than 4 million times the mass of the Sun but only about 17 times wider.

Consider the challenge of capturing a portrait of an entirely dark object only 17 times the width of an ordinary star at a distance of 26,000 light-years. Resolving an image of Sagittarius A* is comparable to resolving the image of a piece of fruit on the Moon. To resolve a shadow about 25 million kilometers across and almost a million trillion kilometers away is to resolve the shadow of an object that fills as much of our field of view as a needle at 10 billion kilometers. To resolve such a minuscule image requires a telescope the size of the entire Earth.

Despite the abundance of supermassive black holes in galaxies, all others are too distant to resolve even with a telescope the size of the Earth. There is one exception. M87 is an enormous elliptical galaxy 55 million light-years away that harbors a staggering supermassive black hole billions of times the mass of the Sun. Bigger, but farther, the supermassive black hole in M87 is as small in our sky as Sag A*.

Exploiting large radio telescopes around the globe—relying on the newest, most sophisticated observatories and reviving some that were nearly defunct—the Event Horizon Telescope becomes a composite telescope the size of the Earth. As the planet spins and orbits, the target black holes rise into the field of view of component telescopes around the planet. To render a precise image, the telescopes need to operate as one, which involves sensitive time corrections so that effectively one global eye looks toward the black hole. As the EHT group framed the technological challenge: resolving the image of either Sag A* or M87 is like reading the date on a quarter in San Francisco from New York City.

The EHT is still staring at Sag A*, persistent. In the meantime, the project revealed its data on M87. On April 10, 2019, at the National Press

Club, a press conference erupted in applause and some tears. The image is unmistakable, a dark shadow the size of our solar system, enveloped by a bright beautiful blob. At the reveal, I was overwhelmingly moved by the sensation of sharing the moment with our species. Billions of citizens of humanity paused as I did to take in the sight, all of us bound to this rocky planet, floating with our solar system in an ocean of dispersed celestial bodies, transfixed by the image of an immense supermassive black hole in another galaxy.

Maybe you alone will have the experience of laying eyes on a black hole. Visit near enough to a supermassive black hole and you could see the shadow possibly with no augmentation beyond your eyes, without a telescope the size of the globe. To get to either Sag A* or M87, you will need to boost to near light's speed. Traveling at 0.9999995 times the speed of light—just a fraction of one-millionth slower than the speed of light, so very nearly 300,000 kilometers per second—the trip to Sag A* would take 26 years of your life while Earth experienced the passage of 26,000 years. If you travel at a speed a fraction of a trillionth slower than the speed of light, you could get all the way to M87 in 26 of your years, while Earth endured 55 million years of erosion. The faster you travel, the quicker your experience of the journey and the more likely you are to survive and disembark at your destination still vigorous with youth. Travel at very nearly the speed of light and you can arrive at the most distant black hole on the observable horizon, though you will have to endure the anguish of the loss of anyone who ever will have lived their life on Earth. You will have outlasted the Sun and the entire solar system, the senescent Milky Way unrecognizable.

CHAPTER 9

Evaporation

I hope my recommendations for navigating pure spacetime will stand you in good stead. Take notes in your field guide to share with us should you make your way home or should your transmissions weather the journey without you. I do have to confess that I have not been entirely forthright, though. I have appealed to you to accept the premise that isolated black holes are dark. Much of my advice, still valid I assure you, offers techniques for negotiating that defining feature of black holes. Still, I must admit, there is one thin vapor of a hitch that makes the premise that black holes are dark technically false: Hawking radiation.

Having established that black holes by much evidence exist, we go back to the pure theoretical isolated black holes of our thought experiments.

Even though black holes lumber in space, huge astrophysical objects siphoning cosmic detritus out of the universe, they are also flawless, fundamental gravitational phenomena. In their barest, abstract form, black holes are an ideal place for skirmishes into the single most onerous campaign in the pursuit of a complete portrait of nature.

The natural laws are few in number. There are effectively two kinds of physical law: one for gravity and one for matter. We have equated gravity with spacetime—matter and energy tell spacetime how to curve, and the curves in spacetime tell matter and energy how to move, to paraphrase Wheeler. The mandates of the matter forces that control nuclei and the atom, that are responsible for the nature of light and much of our experience of the world, seem to be intrinsically different from the mandates of spacetime. Still, a list of two is pleasingly brief. From this impressively short list, the entire complexity of the-world-as-we-know-it emerges, obfuscating the elemental simplicity.

The list of fundamental laws is impressively short, but not short enough. There's an aspiration to uncover a single unified theory of everything, the ultimate physical law that unites all into one.

The campaign to reveal a theory of everything is motivated by the compelling suggestion that gravity and the matter forces, despite convincing appearances, are actually different expressions of the same underlying phenomenon.

Quantum mechanics is not one of the fundamental laws. It's a paradigm for the high-energy expression of each of the laws. At high energies, you can probe minute distances with greater surgical precision—consider high-energy X-rays for instance, small enough to drive past the atoms in your skin, small enough to notice that your skin is really mostly empty space. At high energies, matter is known to behave differently than at the lower energies of our common experience. The high-energy behavior exposes that matter is quantized into discrete indivisible units, quanta. We have superbly well-tested quantum descriptions of matter. If we could magnify the workings of the microscopic universe, we would observe a counterintuitive world of restless quanta that seems indefinite, probabilistic, the fundamental existence of things clouded. Our familiar experience of a simple deterministic reality of unambiguous objects is the illusion, a deception made possible because of our poor perception, our blurry vision, our slow reflexes, our limited strength. To achieve such a profound unification of all natural law into one, we need to grasp a quantum description of *gravity,* and that has proven deftly elusive.

To collect clues about quantum gravity, explore the frontier of black holes. The black hole acts as a magnifying glass amplifying the physical processes on the smallest scales and highest known energies in the history of the universe, precisely the domain of quantum mechanics.

Black holes are indeed nothing, but nothing isn't as simple as the term implies. At the quantum level, the vacuum is never as empty as I proffered. The vacuum is actually plagued with quantum possibilities, with stuff that is there and not there, fluctuating into and out of existence. The ambiguity of existence at the quantum level is related to the Heisenberg uncertainty principle, which states that no subatomic particle can be precisely motionless at a specific place. Before uncertainty was discovered, we might have imagined a microscopic particle that could be pinned motionless at a place—hardly a severe demand—like a well-behaved billiard ball. But our imagined particle cannot exist in reality.

Let me try this analogy, with the caveat that as with all analogies it will have flaws. Imagine you play a chord on a guitar. That chord is a superposition of notes. If you play one definite chord, you are not playing one definite note. You could try to reverse the musical analogy. Play one note by superposing chords such that all notes are canceled except for the one that you want. (Consider the technology of noise-canceling headphones.) Then a note is a superposition of chords, just as a chord is a superposition of notes. There can be a chord or a note, but not a sound that is simultaneously a chord and a single note.

Quantum uncertainty arises because, in analogy to the chord versus the note, a quantum particle cannot at once be in a precise place and simultaneously have a precise motion. If a particle is at a specific location, it is in a superposition of motions. The reverse is also true. If the particle is moving at a specific speed, it is in a superposition of locations. The particle has not materialized definitively in space. It is literally neither here nor there, but both. We say that position and velocity are complementary observables.

We used to imagine that there were these things called particles and they had a location and they had some motion. But then we learned that was as silly as saying there was this thing that was simultaneously a chord and a single note. There is no such thing that is simultaneously a chord and a single note. And there is no such thing that is at a precise place with a precise speed. This whole predicament suggests a more general perspective on the quantum world. At the quantum level it is entirely natural and sustainable to be in a superposition of here and there, then and now, fast and slow. This whole clunking reality of stuff just sitting there, being where it is, is a prosaic illusion. All we have to do is look closer and the firm reality disintegrates.

Extrapolating to the extreme, if you cannot say definitively that a particle is exactly there, then the presence or absence of a particle ceases to be clearly definable. Pushing the boundary further, a fundamental physical limit exists to the sheer emptiness of a vacuum. If we cannot say precisely that something is there, we cannot say precisely that something is not there. As a direct consequence of uncertainty, fluctuations in which particles appear and disappear are irrepressible.

Consequently, not even the vacuum can be entirely empty. The vacuum froths with quantum fluctuations of matter. These quantum fluctuations occur even in the room you occupy. But given that we're big and ungainly compared to the microscopic scales of the quantum sea, we simply cannot notice such fine-grained behavior.

There are rules to the fluctuations in the vacuum, conditions that quantum particles must respect as they appear and disappear. Trying a color analogy, suppose you had a bowl of green paint and an alchemical process that extracted two droplets, one blue and one yellow such that they combine to make precisely the right green. In this analogy the vacuum is a specific state, like a specific green color, and only pairs that combine to reproduce that exact green could possibly fluctuate out of that vacuum. A pair of colors can pop out of the vacuum, borrowing from Heisenberg uncertainty, but they recombine in a flash and go back to the green they came from. Ordinarily, the fluctuations of the vacuum are imperceptible. Pairs materialize and dematerialize again too quickly to perceive.

However, if the vacuum is near a black hole event horizon, then the event horizon can irrevocably separate the droplets, and that is the crucial attribute that allows a black hole to pluck real particles from the vacuum while your room does not. In our color analogy, a blue droplet and a yellow droplet can emerge from green, but if the blue falls behind the horizon while the yellow does not, the yellow left outside the black hole cannot return to green without its blue partner. It is freed to live in the universe, a droplet of yellow that seemed to come from nothing, while the blue fell into the hole.

In black hole language, one photon of a quantum-created pair can fall just inside the horizon and the other just outside. The one that fell in cannot ever get back out to recombine with its partner. Without its partner, the photon on the outside cannot return to the vacuum, as it doesn't have the right properties anymore— it's not the right color in our analogy. The photon that did not fall in can break free since the escape velocity outside of the event horizon is less than the speed of light. The black hole has pulled light out of the quantum vacuum permanently, absorbing one photon and letting the other escape to the edge of the universe.

No specifics are encrypted in the light that escapes. In some sense, the creation of the light had nothing to do with the hole, just the nature of the vacuum, and so no features or information about the hole is encoded in that radiation. From near the black hole emanates a featureless spectrum of light, named Hawking radiation after Stephen Hawking, the brash, ailing Oxford physicist who thought these extraordinary, renegade thoughts.

The heat and energy the Hawking radiation carries away comes at the expense of some energy source, since energy is rigorously conserved. The energy of the light that escapes is exactly balanced by the negative energy of the light that falls in. While you might worry that negative energy is not possible, energy itself is relative and depends on whom you ask. From the perspective of an astronaut inside the black hole, the infalling light will have a positive energy. From the perspective of the black hole, the infalling light contributes negatively to, thereby decreasing, the black hole's gravitational energy in the form of its mass. As black holes radiate Hawking particles, they must lose mass. Black holes must evaporate.

Nothing can escape from a black hole, yet black holes emit Hawking radiation and thereby evaporate. It's fantastically counterintuitive and beautifully ironic. The attribute that makes black holes epically dark, the event horizon, is precisely the attribute that defies their essence and allows a black hole to shine with quantum radiation.

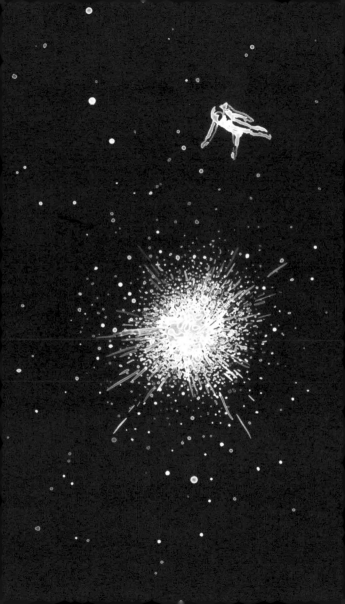

In the context of quantum mechanics, black holes present a genuine physical dilemma. The very existence of the event horizon, a purely gravitational phenomenon, ensures that the Hawking radiation that emerges from the vacuum near the black hole carries no record of the details of the interior of the black hole and no relic of its history. Information that went into the hole never comes out. It cannot be pressed into the heat of the radiation that Hawking proved must leak into space just outside the horizon. Eventually the black hole will explode in Hawking radiation. Nothing will be left, not even the event horizon. The fabric is yanked up and there's nothing behind the curtain. Nothing. Everything and anything that fell in has disappeared.

Black Holes Are Still Dark, Mostly

Hawking radiation is not likely to be important observationally in the lifetime of the human species. All the black holes we have identified are astronomically large and, as before, bigger black holes are milder than smaller black holes. Gravitational acceleration is intrinsically stronger the denser the source, and bigger black holes are less dense than smaller ones. The gravitational energy of smaller black holes is more concentrated and so the gravitational acceleration near the horizon is correspondingly stronger. The greater gravitational acceleration reflects the exaggeration of gravitational forces around small black holes and the consequent probing of high-energy quantum phenomena. The quantum processes driving Hawking radiation are thereby more vivid for tiny event horizons. The bigger the black hole, the fainter the Hawking radiation. The smaller the black hole, the more explosive the Hawking radiation.

Hawking evaporation renders mini black holes both more benign and more dangerous. Suppose you do invent and construct a mini black hole factory in your space laboratory. The microscopic black hole products will not have a chance to absorb your spacecraft or you before evaporating in a flash, within an infinitesimal fraction of a second. But since small black holes do evaporate sensationally near the end, you are effectively setting off lethal firecrackers in your space laboratory, which has its own attendant dangers. The black holes are perhaps more like microscopic dynamite than firecrackers, detonating with the energy of a quarter ton of TNT into potentially fatal explosions releasing harmful X-rays and gamma rays. Create many in a given collision and you have a celebratory display.

If instead you conceive of a means to implode your entire space station, the resultant black hole—smaller than a proton—would erupt with the energy of a nuclear bomb. Self-destruct emergency protocol is never recommended. Employ under only the most dire of circumstances.

You are safe from Hawking radiation around stellar-mass black holes. For a black hole the mass of the Sun, the radiation is so tepid as to be unobservable. The black holes we observe have a Hawking temperature cooler than the light left over from the big bang. These black holes are absorbing more than they are emitting. In the far future, they will become hotter than the ambient universe and will begin the process of evaporation, which takes epically longer than the current age of the universe. You could gaze out the viewports of the space station (before the self-destruct incident) and admire the spectacle, possibly nostalgic if you have admired the glories of the northern lights back on Earth. The dark dead stars will brighten languidly as the event horizons slowly fizzle.

CHAPTER 10

Information

Black hole evaporation reveals unanticipated opportunities for exploration and peculiar misadventures. The Hawking radiation, according to the precepts of gravitational law, may reveal absolutely nothing about the black hole. You and Alice can watch the display from your space station but merely as spectators. No story is relayed to you through the light emanated. There is no valuable information to extract. You could simply kick back and enjoy the rare moment as sheer entertainment.

Alternatively, the precepts of gravity as we currently understand them may falter and the Hawking radiation will necessarily cipher secrets of the interior. Uncertainty is not an invitation for laxity and inattention. You and Alice are impelled to exploit the opportunity to observe the light and scour for information to extract.

Important to our discussion is the very concept of information, essentially everything that can be known about quantum particles and their configurations. Quantum bits of information, qubits, can be bartered, swapped, and reordered, but not destroyed. The original architects of quantum mechanics insisted on this conservation as a philosophical principle worth respecting. They built quantum mechanics to operationally safeguard information.

The consequence of information conservation is reversibility. If information is always protected, then you can predict the future and reconstruct the past. In practice, the reconstruction may be impossible, but in principle, it is physically possible. If your computer catches fire and melts and you mourn lost data, photos, unfinished books, you could enjoy some foolish optimism. In the heat of the fire, in the impact on the molecules in the air, in the carcass of your computer drives, all of that information exists. The information has been hopelessly reordered, but it is physically there and could be reassembled back into a song encrypted in binary, in principle.

Since the event horizon looks the same whether the black hole is made up of my information or yours or computers or databases, the information protected behind the featureless façade of the event horizon must increase as the black hole gets heavier with each gadget consumed. However, the opposite must also be the case. If a black hole loses mass, its information content goes down, but the bits cannot have gone into the Hawking radiation if the event horizon forbids the transmission of all data. The quantum information, the fundamental quantum facts about the material that built up the mass of the black hole that the Hawking radiation then depleted, all of that indestructible information that fell into the black hole has vanished. The universe ends with no hole, no matter in the interior, just the weird debris that radiated during the vaporization process. Once the black hole evaporates, the information previously hidden seems to have actually disappeared.

And there is the catalyst of the crisis in the form of the black hole information-loss paradox. The black hole magically made the information vanish. But information is sacred and cannot vanish.

The architects of quantum mechanics could

be wrong of course. Information might not be sacred. But the consequence would be a terrible kind of unpredictability. A fundamental inability to reconstruct the past from the future is as bad as a fundamental inability to predict the future from the past, which is essentially the entire program of physics. Or, put more simply, physics is devoted to the notion that there is cause and effect and that nature is in principle knowable. If information is not preserved, then nature is ultimately unknowable.

Now, okay, quantum mechanics already undermines some of the most flat-footed determinacy we're used to. To predict the trajectory of the billiard ball we need its velocity *and* its location. Heisenberg surmised that we can have one with unyielding precision or the other, but not both simultaneously. But that's just a problem with quanta like electrons, which we've already argued aren't real in the conventional sense. Inspired by Heisenberg's uncertainty principle, the pioneers of quantum theory pushed uncertainty deeper into the fibers of the physical world. The quantum theorists of the twentieth century suggested that matter was better described by a wave at the quantum level than a particle styled after a billiard ball. A mathematical object called the wave function was proposed. The wave function tells you the probability you will find a manifestation of a quantum particle anywhere in space, the probability you will find that quantum moving at some speed, the probability you will find that quantum in any particular, conceivable state. In the wave function, the idea that chords (a particle's location) are a sum of notes (motions), and vice versa, can be formalized. Under pressure of experiment and exploration, the wave function emerged as more fundamental, more real. The evolution of the wave function is entirely deterministic. The wave function preserves information and its future can be predicted from its past. Its past can be reconstructed from its present. All of the hang-ups scientists previously had about particles—that their paths through the world must be deterministic, that they must really be there in reality, really—are transferred both mathematically and psychologically onto the wave function.

The sanctity of the wave function, the object into which all aspirations for reality and determinism have been invested, has been compromised by the black hole. Worse than the usual uncertainty of quantum mechanics encoded in a firmly determined wave function, the universe ends in a confused superposition of possible wave functions, with the predictability of the future, and the reconstruction of the past, hopelessly lost.

Hawking initiated an insurgence against a fair and reasonable universe. Quantum mechanics is fatally flawed if information genuinely can disappear, as opposed to merely hide. Devout quantum theorists cannot give up information preservation without giving up the entire quantum paradigm, the most precisely measured paradigm in the history of physics. The vast successes of quantum thinking have been too compelling to simply accept such a tragic failing. Information is preserved and no process in the physical universe has the right to destroy it, quantum proponents would testify. Not even a black hole.

By contrast, proponents for relativity were equally certain. A black hole would not let anything go. Not ever. Not even information.

If an obsessive curiosity grips you and Alice, one of you has to consider taking the plunge. If you choose exploration over safety, you could opt to surrender to spacetime and allow the black hole to absorb you while you record your deterioration in the name of science. You and Alice must understand you will not see each other again.

As the solitary astronaut to have explored the interior of a black hole, you alone could resolve the paradox. Although by then Alice will have expired from the burdens of time, for you the crushing agony, pulverized by the tidal forces, will come upon you swiftly. You send messages in vain to no one. But send them anyway. Your epiphanies are forever lost in the catastrophic singularity. But send them, please, your acts of defiance.

The story is revised if this scenario is inaccurate and information escapes. Then when you explore the black hole, you will find no evidence of a true singularity. Saved by quantum gravity, you could uncover the mechanism by which information gets out. Whatever your fate, the mechanism necessarily dismantles you into your constituent pieces to process and release your information. You must record your results quickly, before your disintegration. Your experimental data, the resolution of the paradox that you discovered, that information too will eventually escape in Hawking radiation. If any sentient creatures outside the black hole are driven by curiosity, their telescopes can harvest the light and decode your messages. They could reconstruct the final moments of your experiment, appreciate your historic sacrifice, and document the culmination of your life's work for posterity. Their scientists could take up the charge inspired by your legacy and share your experience of life and death inside a black hole. You will not be forgotten.

In the absence of your communications, a protracted war broke out in the mid-1970s between defenders of general relativity and avengers of quantum mechanics. There were enemy lines, conflicts of ideologies, and wagers placed. The clash between gravity and quantum mechanics would not be waved away, as the next four decades of debate would prove. But its resilience only made the information-loss paradox more exciting, more important. The resolution is not likely to be a fine detail, more a dramatic revision, a sharp turn in the direction of the final theory of everything.

Over many decades, the war would turn in the favor of the quantum advocates, but not readily. From the bloodbath some glorious thoughts would take hold. *Events are not real. Wormholes abound. The world is a hologram.*

CHAPTER 11

Holograms

As in the early days of quantum mechanics and its principles of complementarity, we are faced with two contradictory conditions, both of which we badly want to respect. On the one hand is Einstein's happiest thought. Gravity is weightlessness. The transition into the event horizon of a big black hole should be a comfortable, weightless fall into a shadow, no different than falling into the shadow of the Moon. On the other hand, we have the sanctity of the wave function, which has inherited all of the crucial traits of a sensible physical world, predictability and determinism—the sanctity of information. If we cannot give up on gravity and we cannot give up on quantum mechanics, we can wonder if nature could be taunting us to do the impossible and accept both.

The story of your fall into the black hole can be retold. You are defined by your staggering, vast collection of information, which you can think of as your quantum state. The event horizon poses no obstacle to your descent. You cross that demarcation irrevocably. You believe that you and your information fall into the black hole never to escape. Your death near the center is anguish. Soon after your consciousness is decimated, your qubits are destroyed in the singularity.

Suppose from her space station outside the black hole, Alice (or her descendant) observes a paradoxical scenario. She fastidiously collects the Hawking light emitted and decodes the information to conclude that your qubits were radiated by the hole. This need not be problematic for her if she believes that you never really fell in. To Alice, the event horizon not only slows the pace of your time, the horizon also smears you out across space, thereby amplifying your quantum constituents like a magnifying glass and suspending you there until those bits find escape in the Hawking light.

In this retelling, we are challenged to maintain two mutually exclusive propositions. The information falls into the black hole never to escape *and* the information escapes. Which is it? And here is the radical suggestion: *both.* That is the conjecture. Both happen. It's as though you, or at least your information, has a double. You and your qubits have fallen into the hole, while your double in qubits also escapes the hole. You and Alice don't even agree on your extermination, on basic facts, and your composite reality corrodes even more. But, the designers of the idea say placatingly that no one will ever know, no one will ever be able to observe two contradictory events. You and Alice will never be reunited to tell your sorrowful tales. Alice would recover your qubits outside the black hole like ashes from a cremation and be none the wiser if she leapt in after you to meet her death alone in the interior, your double long since dispensed with by the singularity.

If proven, we have struck pretty close to a proof of the nonexistence of God: there can be no superobserver, no omniscient being. But let's not get carried away. Put that aside. A seed has been planted in a terrain that has proven inhospitable to simple realism. And while that's a lot of tradition to give up just to resolve the information-loss paradox, maybe the payoff will be as great. Undermine notions of reality, but ameliorate the pain with a mind-searing vision of nature.

Holography

To those of us living outside the black hole, our observations of the world, including the black hole and its attendant radiation, are consistent with information accumulating on the near horizon region in a thin quantum layer. The information that we had imagined was contained within its volume, it is hypothesized, is actually encoded on the boundary of the black hole.

While you might think you could pack more information in a volume than you could on any boundary containing that volume, you actually cannot. The volume will not tolerate that much information. We know from mathematical calculations that the information content of the black hole grows as the area of the event horizon grows, independent of the volume within. You can never stuff a black hole with more information than can be encoded on the boundary of the black hole—that is, on the horizon.

The black hole, and please feel welcome to take a beat to accept this conclusion as a consequence of the argument, *is a hologram*—a two-dimensional encryption that projects a

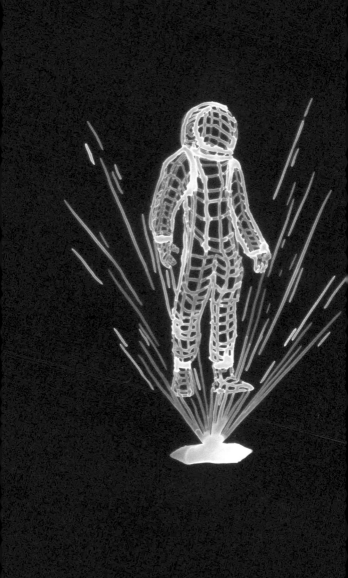

three-dimensional image. All information in the hole is encrypted on the boundary. The three-dimensional interior is an extravagance, a projection, a party trick.

You can never have more information inside any volume in the universe than can be encoded on the boundary of that volume. If you tried, you would make a black hole, and we already know how much information a black hole contains: as much as the event horizon can tolerate. It's not just the black hole that's a hologram. The whole world is a hologram.

Holography is profound and deeply intuitive and simultaneously uncomfortably counterintuitive. The holographic principle remains a conjecture, although a mathematical discovery came close to a proof.

An equivalence was discovered between two apparently very different worlds. Imagine a world with gravity and, as a consequence, with black holes. Players in that universe are confused about the information-loss paradox. That world is effectively in a box. On the boundary of the box resides another world, a microcosm of pure quantum matter with no gravity and so no black holes and therefore no loss of information.

Now, the world in the box has been discovered to be the same exact reality as the world on the edge of the box. An ingenious mathematical dictionary faithfully translates the world with gravity in the box to the world on the boundary without gravity. (This is known in technical jargon as a duality between anti–de Sitter space and conformal field theory.) All of the information contained in the world in the box is faithfully encoded in the world on the edge of the box. There is one reality, one world, which can be described in two ways.

The duality (one world, two descriptions) between two starkly different realities—the worlds do not even occupy the same number of dimensions—provides the first mathematically convincing formulation of the holographic principle. All events played out cinematically inside the box are mere holographic projections of events on the boundary.

You believe yourself to be a three-dimensional astronaut, plummeting through a black hole under the influence of gravity. But there is an equally valid interpretation, equally true. You are really just a deluded hologram, a projection of a two-dimensional reality, an extravagant storyteller with a vivid imagination.

The boundary world is subject solely to quan-

tum principles guiding matter, which, by construction, conserves information. The world inside the boundary is merely a different translation, so it *must* also preserve information. Information cannot be lost in one translation of the same universe if it is not lost in the other. It seems that quantum mechanics has prevailed, information must be preserved, and gravity's stronghold has broken. Although, exactly how the information gets out of the black hole remains a mystery.

The enthusiasm for this stunning duality cannot be underestimated. The conclusion is spectacular and provocative and sweeping and eventually Hawking himself was unable to deny the cogency of the arguments. Although some worry that he did so too soon, Hawking conceded and the war was over. Temporarily. In retrospect, the hostilities merely reached a détente.

Preservation of information refused to waver. The narrative of holography and dualities compelling, all that was left was to tie up the loose ends. But instead, when those ends were tugged, the treaty began to unravel.

. . . .

CHAPTER 12

Firewalls

The black hole information-loss conundrum remains unresolved, the peace shattered in a blazing firewall that even the proponents of the reanimated crisis lit with regret.

Entanglement

The firewall controversy requires a little extra machinery: quantum entanglement. Quantum entanglement is wonderfully weird. Let's start with something that's not weird but instead rather ordinary, and then see how the scenario unfolds when we try the same sport with quantum players.

Suppose you and Alice break a wishbone. There are two possible outcomes of the contest. You have the big piece, Alice has the small piece, or you have the small piece, Alice has the big piece. Neither of you looks at the outcome of the competition. You each put your piece in your pocket. You travel to Andromeda and you look at your shard of bone. You see that you have the small piece. You instantly know that Alice must have the big piece. There's nothing spooky about this sequence in the least. The information was there in your pocket when you left the dinner table and traveled with you to Andromeda. Back on Earth, Alice looks at her piece of bone and sees she has the winning splinter. She instantly knows you have the losing fragment. Her knowledge and the result of her experiment, the experiment being a glance at her piece of the bone, is entirely independent of your knowledge and the result of your experiment, your experiment being a glance at your piece of the bone. The outcome of the observation was determined back at the dinner table.

The quantum analogue of the wishbone experiment is so much stranger. Suppose instead of a wishbone, you and Alice each catch a quantum particle that fluctuates out of the vacuum as described in the context of Hawking radiation. The particles always come in pairs, which we call Hawking pairs. Just like the broken bone must fragment into two pieces that fit back together to make the original wishbone, the pairs need to complement each other to reflect the original whole from which they came. In this case, it's the vacuum that split into the pair of particles. Suppose, in the context of our previous color analogy, that the vacuum is again like a bowl of green paint and there exists the same alchemical process to separate a yellow droplet. Then there must be a very specific blue droplet that when combined with the yellow had better give the green paint you started with. In order to return precisely the right green of the vacuum if recombined, there are two possibilities for the pair: the droplet you caught is yellow and the one Alice has is blue, or the droplet you have is blue and Alice has yellow.

Here's the dramatic difference from the wishbone contest. Like the sound that can be a superposition of chords, your particle can be in a superposition of colors. But the particle you have and the particle Alice has, if combined, must return to the green they came from. Your particles cannot separately be in a superposition of colors. They have to be in an *entangled* superposition of colors: your droplet blue, Alice's yellow *and* simultaneously your droplet yellow, Alice's blue. If you separate the pair

gingerly enough, they will remain entangled in
this superposition even as you each put one in
your pocket and you leave for Andromeda. The
information on the winning contestant has not
yet been determined. (Let's say you win if you
have blue.) The information isn't set. The particle in your pocket is not yet blue or yellow. It
is in an entangled superposition of blue (while
Alice's particle is yellow) *and* yellow (while
Alice's particle is blue).

You get to Andromeda, you look at the particle in your pocket, thereby disturbing it and
destroying the superposition. When you observe
your particle, the vast numbers of quanta in your
experimental apparatus and in you are unable to
maintain the delicate superposition. You force
the particle to materialize into one distinct state,
either blue or yellow. It is blue. You have won.
You have also received information faster than
the speed of light. Instantaneously, you know
Alice has the yellow particle. Instantaneously,
your disruption of the entangled superposition
forces Alice's particle millions of light-years
away to assume the yellow state. It does so faster
than the speeding light could get from Andromeda to Earth.

Alice does not however instantaneously come

to know you have forced her particle to the yellow state. She could look and discover her particle was yellow, but she always had fifty-fifty odds of a yellow droplet, and so her losing result provides no intrinsic information about what you have done. Alice will understand that you must possess blue but has no way to know that you have performed your measurement yet. For all she knows, she was the one who forced the pair to assume one combination over the other. Alice has no way to know what you know. If you want Alice to know you looked at your droplet and won with blue, you would have to resort to conventional subluminal means, like a gloating letter sent through intergalactic post. Then she could go look at the droplet in her possession and verify your claim, if she hasn't done so already. It will indeed be yellow.

Einstein famously thought up this quandary with Boris Podolsky and Nathan Rosen. He called the result *spukhafte Fernwirkung*, which is usually translated from the German as "spooky action at a distance" (or sometimes "ghostly" in lieu of "spooky"), and waved the example around as evidence of purported flaws in quantum mechanics. Summary of grievances against quantum mechanics: For one, quantum mechanics does not respect relativity and does not respect light's speed limit. Quantum mechanics instead allows for spooky action at a distance. For another, quantum mechanics has violated local realism. Quantum observables in the experiment did not have a definite value, and so were not real, at least not in the conventional sense, and not until you measured them.

Einstein intended to mock the absurdity of quantum theory. You can entangle particles and delicately separate them to preserve the entanglement. You can then communicate the information about their quantum states faster than the speed of light. It's absurd, but by all observations and laboratory experiments, it is also true. The strict rules of entanglement imposed on evaporating black holes inform the next push in the climb toward a consistent quantum description of gravity.

Firewalls

The vacuum is the emptiest state possible, which, as we've already established, still froths with virtual particles, none of which are detectable ordinarily. A pair of virtual particles created as allowed by uncertainty must preserve the state of the vacuum. If one is blue, the other is yellow in order to conserve the state of the vacuum, which is metaphorically green. Consequently, Hawking pairs must be entangled. The photon that shines to the outside must be created in an entangled state with the photon that fell in the black hole. Since the horizon is a hopeless divider, the pair is entangled forever.

Entanglement must be monogamous. Each particle in a Hawking pair is maximally entangled with its partner. Every detail of its quantum state is bound to a detail of its partner to ensure that together they respect the conditions of the vacuum, as in the color analogy. All of a Hawking photon's information is already bound up with its partner's information. There is no information free to tangle together with anyone else's. If Lucy is a Hawking photon that escapes, and Sam is her counterpoint that falls into the black hole, then Lucy must be faithfully entangled with Sam. If Lucy and Sam are maximally entangled, they are so entangled monogamously.

Fine. Here is the problem. Suppose that relativity loses and quantum mechanics triumphs such that somehow the Hawking radiation carries all of the information out over the course of the black hole evaporation. If you collect the black hole's Hawking radiation over the eons, you should be able to reconstruct the story of what fell into the black hole. In order to tell you that story accurately, the Hawking radiation must contain information and must be carefully apprised of the information freed from the black hole earlier. The way in which the Hawking radiation is apprised of the information is through entanglement. The Hawking radiation that escapes at later times must be entangled with Hawking radiation that escapes at earlier times. But we've already insisted that Hawking pairs are entangled with each other so they cannot simultaneously be entangled with anything else. That would be polygamous. If Lucy is a Hawking photon that escapes, she is faithfully entangled with Sam, her Hawking partner who fell in, and cannot also be involved with Pam, who escaped trillions and trillions and trillions of years ago. Again, that would be polygamous.

For emphasis, there is a contradiction between the entanglement in the radiation necessary to preserve information and the entanglement necessary to have an empty vacuum near the horizon. You cannot have both.

The contradiction can be sharpened in a provocative restating of the crisis. If you believe that information must be preserved, you are most concerned that the late and early radiation tell the beginning, middle, and end of the story of information. Consider the implications of preserving at all costs the monogamous entanglement between early and late radiation required. Since polygamy is forbidden, we would have to give up the belief that the Hawking pairs are entangled when radiated. In the color analogy, maybe the divided yellow droplet cannot be entangled with the other droplet, so the other droplet might not in fact be blue. But then the original conditions from which the pair sprang must not be the vacuum—the original bowl of paint can't be that original green. If there's no correlation between the colors of the droplet pairs, then the original bowl of paint must be a splotchy mess of color. If we give up on our previous belief that the Hawking pairs are entangled—if they are not entangled—then they must not have come from the vacuum, but rather from a splotchy mess. The alternative to emptiness is fullness. If the event horizon isn't empty, maybe it's actually full of a fiery wall of stuff. Maybe instead of nothing there is an incandescent event horizon, a blazing hot firewall.

If there is a firewall, then no one ever gets to the interior of the black hole, there is no interior.

The black hole ends at that firewall around the event horizon and any attempt to trespass results in unambiguous cremation.

Black holes are nothing. I spent half a book convincing you of that. According to relativity, we all would have sworn that you would experience no drama as you fell across the event horizon, as the new lexicon would have it. If you fell across the horizon in an inert spacecraft, your experience would be identical to that in a falling elevator cab. You would feel weightless and any experiments you carry out would be the same as if you fell freely in empty space. As you cross the horizon, nothing extraordinary should happen.

If instead there is drama on the horizon and you are incinerated in a firewall, then relativity and the deeply loved equivalence principle are wrong about black holes and who knows what else. Or quantum mechanics is not trustworthy near the event horizon and who knows where else. The physics community faces a genuine dilemma that cannot be ignored.

Truthfully, there is little love for firewalls, and theorists are hard at work to banish them from the world. Firewalls very probably do not exist. The thought experiment that imagined them, however, discovered a quandary begging for a resolution. The important impact of the firewall provocation was to elucidate salient details that will lead to progress in understanding the quantum nature of reality. The black hole is offering us clues about the truth, and if we follow those clues, we explore territory that leaves convention behind. This is where we find ourselves at the time of this writing, lost in thought at the edge of the black hole horizon.

One idea proposed to restore monogamy without firewalls—and meet all the demands to resolve the information-loss paradox—is so brilliantly crazy it deserves mention: hypothetical wormholes connect the Hawking radiation we call Sam who fell into the hole and the Hawking radiation that already escaped whom we call Pam outside the hole so that Sam and Pam are actually the same entity. Then Lucy can be entangled with both Sam and Pam and remain truly monogamous, entangled as she is with only one partner that lives on both sides of the black hole with the help of a wormhole. Then information could escape and there would be no need to invoke a firewall.

Pushing this wild suggestion to an extreme, there may be no unique interior to the black hole. The interior only exists through the connection provided by wormholes to entangled qubits on the outside. When you fall inside the black hole, wormholes ensure that you are also outside the black hole. Your information falls in but also escapes. You will perish. But your information will survive beyond the hole and with the continuance of your qubits, the physical potential to reconstruct your body and your thoughts and your memories—you—also survives. The extermination of your information was a ruse, your own death reversible.

A fascinating extrapolation hints that the event horizon itself emerges as a consequence of quantum-negotiating wormholes. A network of wormholes knit between inside and outside could conceivably create the event horizon, as though the boundary merely emerges from a tangle of sewn thread. Like embroidery whose coherence on closer inspection deteriorates into stitches, there would be no independent event horizon and therefore no autonomous black hole, as though the black hole itself is an illusion that is impossible to maintain under scrutiny.

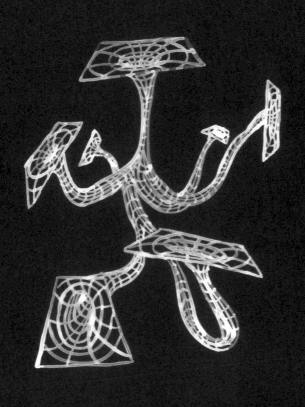

CHAPTER 13

Exit

Black holes are our history and our future. Black holes may have precipitated out of the primordial soup of a baby universe first and may inhabit the dying universe last. The heaviest individuals in our portrait of fundamental particles, they may have been created in the big bang in the earliest, briefest moments of our origin story. Living in a bath of ingredients that are too hot to settle into a recognizable state of matter, primordial black holes would explode and vanish in an early universe that rapidly became too cool to be able to attain again the drastic energies required to re-create them.

Bigger black holes could arise later, though still within the first minutes of this universe, as the primordial soup sloshes and transitions through a sequence of stages, like a developing embryonic world acquiring the specific features of the cosmos we observe. Millions of years later, possibly collapsing directly out of the cooling debris, supermassive black holes could form, skipping stars entirely. Big black holes have more mass, but that mass is concentrated in correspondingly larger horizons. The larger horizon wins in the competition, and bigger black holes are in some sense easier to make, collapsing out of clouds in the early universe that are significantly less dense than a dead star. A supermassive black hole could form out of a fog no denser than air. Once these titanic black holes are born, galaxies and clusters of galaxies would rally around them, with generations of stars living out their life cycles to populate those galaxies with billions of stellar-mass black holes.

Supermassive black holes shape the largest cosmic structures that we observe. They sculpt the galaxies in which they reside, controlling winds and jets that define the sizes and morphologies of the islands of stars. They are agents in the elaborate chronology that made this habitable planet in this Milky Way in this Virgo supercluster in this universe in a possibly infinite multiverse.

Our megalith is Sag A*, the supermassive

black hole that articulates the center of our galaxy. The solar system, along with our entire spiral galaxy, orbits Sag A*. Carried by the Earth and bound to the Moon, we people orbit the Sun, which is 150 million kilometers away, traveling at a speed of nearly 30 kilometers every second to execute a near circle 940 million kilometers long in the span of one year. Our entire solar mechanism orbits Sag A*, which is about 25 hundred thousand trillion kilometers away, traveling at a speed of 230 kilometers every second to execute one orbit in about 230 million years. In galactic years, the Sun is about 20.

The Sun once earned the accolade of center of our planetary system, which was thought to be the center of the universe. And then the Sun was deposed. Sag A* reigns as the ultimate galactic protagonist, orbited by essentially every solar system in the Milky Way, by globular clusters—strange dense globs of erratically moving stars—and by an aptly named halo of invisible dark matter.

The solar system's motion is something of a spectacle, something to provoke the imagination. The Sun and plasma winds and planets and the multifarious ample moons and striated rings and Jupiter's red-raw stormy eye and all the human-made satellites—flicked, shiny lost coins forever falling in the harsh quiet—and Pluto, champion of the hundreds of dwarf planets, and those planets that shattered into neurotic asteroids and interplanetary ice, rocks, fog, magnetic lines . . . all traveling together at a daunting speed exceeding 200 kilometers every second—every few breaths—in the pinwheel spin of our galaxy around an emptiness more massive than 4 million Suns. All the elements of our solar system spinning and orbiting, the original orrery, in mechanical glory like an arcane, wildly ambitious timepiece while we stagger around on one rough gear, oblivious.

We will slowly fall into the Sun and the Sun will slowly fall into that supermassive black hole, along with everything else in the galaxy. But this will take so long that we will collide with the neighboring Andromeda galaxy eons earlier and possibly be tossed out of the Milky Way. For a time, Andromeda will approach and loom large and pale in our view of the sky, threatening to slice through the Milky Way. On impact, we might be passed to Andromeda or thrown near its central supermassive black hole, a thousand times heavier than ours. The stars too sparse to collide, the galaxies will pass through each other with very few direct stellar hits, snowplowing interstellar gases together to glow hot and sore, while the dark matter sails through. Our entire solar system will travel in unison undisrupted by the multiple collisions occurring over billions of years. The two central black holes will merge and one galaxy will coalesce from the wreckage.

When we contemplate the shadow of our local supermassive black hole, we are contemplating our future. That is where our data, our scraps of quantum information, may end up. Look in the direction of Sagittarius as we imperceptibly fall toward our galactic center. Given a near eternity, all of us who have ever lived here on Earth or ever will live, vaporized by our dying Sun into our fundamental elements, will fall into a supermassive black hole at the center of our merged galaxies, as will every other star system, scrap of galactic debris, the entire halo of dark matter. Everything will wash down the central vortex, flashing spectacularly bright, the last desperate blasts of concentrated light in the cosmos, until all vanishes in a darkening silent storm in spacetime.

One day, although it's quaint to think of that moment in the epic future in terms of days, when the universe is nearing the end of its metabolic life, the cosmos may be empty except for black holes, and those black holes will evaporate, likely surrendering the information hoarded, the current controversy resolvable if only there were somebody left to play witness. Our quantum bits possibly will reside simultaneously inside the black hole and outside the black hole. Linked by wormholes, we will be our own clones in two places at once. When all of our information is finally released from the fading event horizon, it will be miserably disordered. Illegible.

Ultimately, there only ever was information. This story of our beginning, our evolution, our ambitions to know, our presence here, will be strewn in an unreadable form no longer registering time, our history effectively erased.

In the end, there is no surviving black holes.

A NOTE ABOUT THE AUTHOR

Janna Levin is a professor of physics and astronomy at Barnard College of Columbia University. She is also director of sciences at Pioneer Works, a center for arts and sciences in Brooklyn. Her previous books include *Black Hole Blues and Other Songs from Outer Space*, *How the Universe Got Its Spots*, and a novel, *A Madman Dreams of Turing Machines*, which won the PEN/Robert W. Bingham Prize. She was recently named a Guggenheim Fellow. She lives in New York City.

A NOTE ABOUT THE ARTIST

Lia Halloran is an artist who often draws from concepts and the history of astronomy. She developed her love of science beginning with her first job at the age of fifteen at the Exploratorium in San Francisco dissecting cow eyes and doing laser demonstrations. She is a painter and photographer living in Los Angeles, represented in Los Angeles by Luis De Jesus Los Angeles, and is an associate professor of art at Chapman University.

A NOTE ON THE TYPE

This book was set in Garamond, a typeface originally designed by the famous Parisian type cutter Claude Garamond (ca. 1480–1561). This version of Garamond was modeled on a 1592 specimen sheet from the Egenolff-Berner foundry, which was produced from types thought to have been brought to Frankfurt by Jacques Sabon (d. 1580).

Typeset by North Market Street Graphics,
Lancaster, Pennsylvania

Printed and bound by Friesens,
Altona, Manitoba

Designed by Betty Lew